SCIENCE WITH CHILDREN

SCIENCE WITH CHILDREN

Doris A. Trojcak

School of Education
University of Missouri—St. Louis

McGraw-Hill Book Company

New York   St. Louis   San Francisco   Auckland   Bogotá   Düsseldorf
Johannesburg   London   Madrid   Mexico   Montreal   New Delhi
Panama   Paris   São Paulo   Singapore   Sydney   Tokyo   Toronto

This book was set in Souvenir Light by Black Dot, Inc.
The editors were Eric M. Munson, William A. Talkington, Jeannine Siliotta, and Susan Gamer;
the designer was Joan E. O'Connor;
the production supervisor was Leroy A. Young.
The drawings were done by J & R Services, Inc.
Von Hoffmann Press, Inc., was printer and binder.

SCIENCE WITH CHILDREN

1 2 3 4 5 6 7 8 9 0 V H V H 7 8 3 2 1 0 9 8
Library of Congress Cataloging in Publication Data

Trojcak, Doris A
  Science with children.

  Includes index.
  1. Science—Study and teaching (Elementary)
I. Title.
LB1585.T76   372.3'5'044   78-25949
ISBN 0-07-065217-1

To my parents

CONTENTS

PREFACE

Beginnings are seldom easy. In most beginnings, there are usually more unknowns than knowns and far more questions than answers. As you read these words, you may be wondering about me; as I wrote them, I wondered about you. I could only speculate about you; but I will tell you something about myself, because I strongly believe that the teaching and learning process is a very special interpersonal relationship.

In a classroom, laboratory, or field setting, interpersonal communication is—or at least can be—relatively easy. But when the mode of communication is the printed word, interpersonal processes become more difficult. At the outset, a great deal of trust is needed. I have to trust my judgment in selecting the best combinations of words for you. (Someone once told me that it's easy to talk in circles; but when one writes, it must be in straight lines.) You, in turn, must also have faith that these printed words are worth considering; then you must decide to accept, reject, or modify them in your own unique ways. If you know something about me, it may be easier to establish trust; and the following chapters may make more sense to you.

I've always liked discovering the outdoors much better than being indoors. I was a kindergarten dropout. I have only two memories of classroom science in the elementary grades: one teacher referred to our science test fairly often—she would tell the children in the back of the room to sit on it if they couldn't see the chalkboard; another teacher had us press leaves in our science book, since they would be undisturbed there. My "formal" introduction to science began in the seventh grade, and I attacked it like a starving child turned loose in a candy store. In high school, science-fair projects and the opportunity to work for the Illinois State Museum and later in the laboratory at our local hospital were some of my favorite extracurricular activities.

As a beginning elementary teacher, I was very much interested in both teaching and learning, erroneously believing that there was a simple causal link between the two. I can well remember when I discovered that there was not. It was as if I had been a child parading around with a huge balloon on a thin stick, and suddenly someone had popped the balloon. For quite some time I wondered what could I now do with the seemingly useless stick of traditional training in education: the simple formula that teaching equals telling or showing and that listening, watching, or reading equals learning. Gradually, however, I began to realize the importance of one of my father's favorite sayings: "If you have the right tools and know how to use them, you can do just about any job." That simple idea has served me well in my professional career as a science educator. For quite some time, I've been seeking, studying, and testing an assortment of ideas, tools, and skills to use with children to make science more meaningful for them. I hope that I have now assembled them in a way that will help you do a better job of science with children.

Acknowledgments

The "Acknowledgments" section of a book represents one of the few times when one can publicly thank others for their help. Often I've read this section in other people's books simply out of curiosity—as you may be doing now. These most personal messages are difficult to communicate. But although the people I mention will be unknown to you, I hope I can somehow convey their importance to me and their presence in this book. To a great extent we are what we take in, and these people are certainly a part of me and my efforts.

Throughout my professional life, many students, colleagues, friends, and associates have helped me to grow personally and professionally. I thank them. But I give my parents the biggest thanks and dedicate this book to them. Their caring has been as constant as any of the laws of science and as generous as any daughter could ever hope for. Special thanks, too, go to Marcia, for her interest and encouragement; to Lois, Carol Ann, Geri, Ceil, Helen, Jim, Ron, Hans, and Jeanne; and especially to Don for reading and reacting to the manuscript and to Sylvia for typing and retyping it. Thank you, McGraw-Hill, for your confidence—especially for Steve's original interest, Jean's patience, Jeannine's thorough work, and Eric's and Susan's guidance.

Doris A. Trojcak

INTRODUCTION

We each perceive uniquely: what is beautiful to one person may be ugly to another; what one person considers important another may consider trivial. Differences in perception are usually more helpful than harmful. They provide the bases for a great variety of approaches to life in general and to education in particular. How boring it would be if there were only one way of seeing, or only one right answer to all questions!

Some books on science education deal strictly with the content of science; some stress the thinking skills used by scientists; some describe teaching techniques and curricula. Some are narrow in their coverage; others are quite broad. It would be unreasonable to expect any one text to meet all the needs of science educators. But it is reasonable to look at the reasons why an approach was selected, and at the basis for the contents of a book.

It's impossible to remember how many times I've asked myself and others what educators need to help children learn (and enjoy learning) science. The answers have tended to cluster around three general areas. It seems to me that teachers need to be able to (1) comprehend the basic notion of science and how children go about learning science, (2) help children acquire the sorts of skills that will enable them to make meaningful discoveries about the natural world, and (3) develop the instructional skills necessary to achieve the first two goals. These three areas are reflected in the three major divisions of this book: Part One, Science and Children; Part Two, Science for Children; and Part Three, Science with Children.

Some ideas for establishing a foundation are presented in Part One. Chapter 1 gives you an opportunity to examine your understanding of science in general as well as some of the problems associated with presenting science to children and some reasons for its importance. Chapters 2 and 3 deal with Piaget and his ideas on how children learn. In my professional life, I've studied the works of many of the learning and developmental psychologists and tried to apply their basic ideas. Those of Piaget have made the most sense to me, and they therefore permeate much of this book. Ultimately, however, only you can decide on their importance to your own work in science with children.

Some of the major skills of science—the thinking operations in general—are presented in the numerous activities of Part Two. These include observing and measuring metrically (Chapter 4); distinguishing spatial and temporal relationships, classifying, and communicating (Chapter 5); inferring, predicting, and hypothesizing (Chapter 6); and controlling variables, interpreting data, and defining operationally (Chapter 7). Part Two builds on the foundation that was established in Part One.

A structure is seldom complete which consists of only a foundation and a framework. You will need more than knowledge about science and children, or science for children, to function in the classroom. Many opportunities to develop and apply some of the basic instructional skills are presented in Part Three. These involve observing classroom interactions (Chapter 8), analyzing and interpreting elementary science curricula (Chapter 9), and, finally, testing decision-making skills (Chapter 10). The significance of Part Three will become more apparent to you as you gradually see how the instructional skills needed to do science with children are related to and supported by the thinking skills of science.

Throughout the development of this book, I was faced with important decisions. What should be included or omitted? When is an explanation sufficient, insufficient, or superfluous? What should be explicit? What should be implicit? My primary tasks were to design the most suitable invitations to learning, to help you assess your present abilities, and to encourage you to interact with new ideas and situations. Your tasks are to select, act upon, and internalize what is most relevant to you. Since it is impossible for me to anticipate the needs of each reader, no specific objectives are identified at the beginning of each chapter. The "Encounters" at the end of each chapter, however, can help you assess your learning. When none of these seem to fit your needs or interests, I hope you will invent your own or modify any of those given. The "Suggested Readings" are additional ways to enhance your learning. I provide some structure and suggestions, then, but it will be your responsibility to select what will best meet your needs.

Throughout this book, I am working from certain premises which I strongly hope you share or at least are willing to consider.

First, I hope you have a genuine interest in helping children with science. Second, I hope you are willing to treat science as a basic along with the three Rs. Third, it will be helpful if you have at least some real interest in science and if there are some smoldering remains of your childhood curiosity about the natural world that could once again be ignited. (You're never too old to hug a tree or learn to know a waterfall by standing in one.) Fourth, I assume that you have a sincere commitment to the profession of education, that you have seriously answered questions like "Why do *you* want to be a teacher?" or "You're *still* teaching?" Fifth, I hope you have an indomitable willingness to work at becoming the best science educator you can be.

It would be foolish for either you or me to believe that this book will automatically transform you into the best possible elementary school science teacher. There are surely other premises and other content worth considering, and personal qualities which only you can develop: concern, openness, patience, trust, awareness, inquisitiveness; and the ability to know yourself and others, to give and receive, to risk and to care. I hope that your interactions with this book will result in both personal and professional growth and will help you to continuously improve your ability to do science with children.

PART ONE

IT'S

extremely risky to embark upon any enterprise without having at least a general notion of what's involved. Presenting science encounters to children is basically an interpersonal relationship centered on interactions with the natural world. Selection of *what*, *how*, *when*, *where*, and *with whom* stems from the process most often used in all of education—namely, the decision-making process.

As a science educator, you will be faced with countless decisions related to science interactions with and for children. If your decisions are to be made thoughtfully and effectively, you must have some basis for your choices. The purpose of this section is to help you begin to recognize some possible bases for your future decisions. The three chapters in this section will enable you to focus on the most basic considerations: science and children. They provide the framework for the remainder of the book. Your view toward science, along with your understanding of children, can become the foundation for your decisions regarding science for children and, eventually, your involvement in science with children.

SCIENCE AND CHILDREN
ESTABLISHING A FOUNDATION

CHAPTER 1
SOME PROBLEMS AND SOME HOPES

It would be a joy to begin as an optimist, by reporting the slogans on bumper stickers found on teachers' cars in elementary school parking lots across the country: "STP = SCIENCE—TOP PRIORITY," for example, and "FORWARD WITH BASICS THROUGH SCIENCE," and "WANTED—MORE SCIENTIFIC LITERACY." But I must begin as a realist and admit that such statements are hopes for the near future rather than realities of the present. There are many problems related to the teaching and learning of elementary school science. They must be identified and analyzed if we can ever hope to overcome them or at least lessen them.

Why Is Science a Problem for Many Elementary Teachers?

During the first class meeting of a course in science education, I like to ask my students, "What would you like to learn about teaching science to children?" Often the result is blank stares, or eyes suddenly gazing at the floor or ceiling, or a rapid increase in doodling. Such nonverbal reactions usually imply, "You're supposed to teach us, so stop these delay tactics and get on with your job." But invariably, if I have the patience to wait for comments, the first will probably be something like, "What makes teaching science so difficult? I'm scared." The first time I received a comment like that, I was somewhat shocked. But I've come to respect the question, and the reaction that prompts it, as legitimate and sincere. Teaching science is a difficult, unnerving task to a great many people, both those who are preparing for the profession and those already in the ranks. Why is this so?

There are probably a multitude of reasons. But I would like to center on just four broad problems. It has been my experience that many preservice and in-service teachers have the following difficulties:

1. An inadequate understanding of science, scientists, or both
2. An inadequate understanding of how children learn science
3. A feeling of being out of touch with contemporary teaching and learning of science at the elementary level
4. Generally, a fear of venturing into the relatively unknown

Some Problems of Understanding Science and Scientists

If you were to try to define the word *science* on the basis of your own experiences, you might find the task rather difficult. There are numerous books you could read just on the meaning of science in general. You could also interview dozens of scientists; and you would soon discover a vast variety of interpretations and definitions. But eventually you would recognize some basic agreements. One point of agreement is the dual nature of science, as *product* and *process*.

As *product*, science is an organized, systematized body of verified knowledge about the natural world. It is the record of what has been discovered about order in the universe, the relationship of matter and energy, the interdependence among organisms, and the interactions of organisms with environmental conditions. It is the study of causes and effects, beginning with observations and leading to generalizations, theories, and eventually laws.

As *process*, science is exploring, searching, discovering—the vast variety of thought processes for acquiring knowledge, beginning with observing and culminating in experimenting (Figure 1-1). It includes the continuous process of verification whereby new findings become the bases for additional predictions.

Perhaps you have been exposed only to science as product, as a

Figure 1 1 As *process*, science is exploring, searching, and discovering. (Photograph by Emily Richard.)

storehouse of knowledge, and have found the scene overwhelming, confusing, or even frustrating. Were you to see a playback of all the science content that had been "taught" to you during your past classroom education, would you be amazed or dismayed at what you've retained? When did you ever experience science as a human enterprise, as your own imaginative, exploratory quest for answers to questions stemming from your observations of inconsistencies? Perhaps you have never had the opportunity to experience science as process—the "doing" aspects of science. If so, you might know science only as a noun; but science should be considered as both a noun and a verb. In fact, science is a combination of nouns and verbs making incomplete sentences. The nouns are any things or events which can (at least theoretically) be observed, from the smallest component of an atom to the scope of the universe. The verbs are the thinking processes like comparing, measuring, communicating, classifying, identifying space and time relationships, inferring, predicting, hypothesizing, and so on. The incomplete sentences are the ongoing, ever-evolving "truths" or best hunches of an Aristotle, a Newton, an Einstein, a student of yours, and even yourself.

It is this dual nature of science, as product and process, that I hope to convey in this book. The title *Science with Children* is intended to be both *declarative* and *imperative*. The book is designed to help you begin to present science as noun, as subject matter, to children so that they can better understand the natural world. It is also an appeal to you to *do* science with children, to help them become actively involved in discovery. I hope that not only will the word *science* become less of a problem, but your understanding of scientists (as well as that of your students) will also improve.

I was once invited to "lecture" to about eighty first-graders at a university lab school. The designated topic was "What Is Science?" I decided that it would be inappropriate to begin directly with the concept "science," so instead I asked if anyone could describe a scientist. Immediately a hand shot up, and a boy on one side of the room confidently and briskly announced, "Scientists are people who make bombs." Then a girl on the other side of the room added, "And those bombs kill people." After I had recovered from my own state of shellshock, we continued the discussion, in which more stereotypes were elicited—the chemist in the shining white lab coat surrounded by an assortment of bubbling, smoking test tubes; the runner with the butterfly net waving overhead; the nurse peering through a microscope; the researcher surrounded by piles of books; and, of course, at least one chalkboard filled with strange symbols. (Children's ideas about scientists are also revealed in their drawings; see Figure 1-2.)

Figure 1-2 Children's images of scientists are often clearly revealed in their drawings.

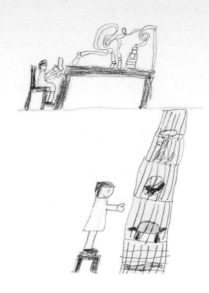

Adults, too, have stereotyped conceptions of scientists. Some think of scientists as dispassionate, pompous, impersonal—almost like robots. Others see them as saviors of the world and solvers of all its problems ("It's 'scientifically tested,' so it's got to be good." "Better living through chemistry.") Rarely is either description true of any particular scientist, and neither is true of scientists in general. But scientists do tend to be different from nonscientists in certain ways. They tend to be more analytical, critical, and objective. Those who have devoted their professional lives to science are more curious and ask more questions. They tend to be more cautious and patient, willing to wait for the evidence and to hold their judgments in abeyance rather than jumping to conclusions. Generally, a scientist would prefer one small proved fact to a deluge of unproved opinions.

Don't let these positive characteristics of scientists alienate you because they might seem to be unattainable or idealistic. You probably possess similar characteristics to a greater extent than you realize. Think about the many times each day that you analyze information or observations, become bothered enough to ask questions, weigh the pros and cons of a situation, and then operate as best you can on the basis of your tentative conclusions.

It's extremely important for you to analyze your view toward science and scientists, because your personal viewpoint will greatly influence how you present science to your students. It will also affect what you expect and accept from them. Will your viewpoint be narrow or broad, accurate or inaccurate? How will you help your students understand the following important distinctions?

The *means* of science are the processes, the thinking skills, or the operations performed in searching for and acquiring knowledge.

The *ends* of science are the products, the actual knowledge that is collected or discovered.

The *guiding principle* or the ethics of science is empiricism: that is, observation and experimentation resulting in findings which can be replicated by others, are based on probability, and are, therefore, never final but always tentative.

The *use* of science is technology, or the application of the products of science (which can range from a toaster to a hydrogen bomb).

The *controllers of science* are scientists, guided by the objective ethics of empiricism.

The *controllers of technology* are you and I—society in the broad sense (of which your students will eventually become members), guided by a subjective value system.

Science can stand alone and is neutral. In themselves, the process of science and the resulting discoveries are objective. But science cannot be separated from technology, which tends to be far more subjective. Its effects can be either positive or negative—depending on how people decide to use it. Again, your understanding of these characteristics and distinctions will influence your science with children. Will your teaching reflect science as a static body of knowledge consisting of the "one right answer" to be memorized or as a constantly expanding, dynamic search for the best hunches? Will children leave your class with the feeling that they "know all there is to know" about plants, or animals, or electricity—or with the realization that there are many more questions yet to be answered?

Some Problems of Understanding How Children Learn Science

This topic will be examined more thoroughly in Chapter 2, Children and Learning, but some preliminary considerations might be helpful here. It has been quite some time since most of us were children. You might have some difficulty remembering a science class during your days in elementary school. Unless you have raised your own children and systematically observed their growth in understanding science, or have done some comparative classroom science teaching, or have thoroughly studied learning psychology and science education, you may feel that your understanding of how children learn science is inadequate. Let me at least plant the seed of understanding here; you can watch it germinate as you progress through the book: Children learn science best by interacting with or acting upon the natural world. Children do not learn science by "looking up" and memorizing meaningless terms. Although they may be fascinated by watching someone perform fabulous feats of science, this type of "learning" leans more toward their favorite explanation of the unexplainable— "It's all magic."

Just as your view of science influences how you present science to children, so does your understanding of children themselves. The younger the child, the greater the problem. Why is this so? It has to do with the greater distance or differences between you and the child (Figure 1-3). Can you remember your thinking processes and the levels of communication between you and your teachers in high school? In junior high? In the sixth grade? In the fifth? What happens as you go farther back? You may remember some specific details of your early years in school, but were you really conscious of how you learned or what kind of thinking processes were going on in your head? It would be most unusual if you were. The ability to understand how children, especially young children, learn science is not easy to achieve, especially for those who lack experience in this area. I hope that Chapters 2 and 3 will give you some valuable insight and lessen this problem.

Some Problems of Teaching and Learning Science

The difficulties of teaching and learning science have become greater as science has increasingly been relegated to a lower status, and been allotted less time, in the elementary school curriculum. In many of the elementary schools I have

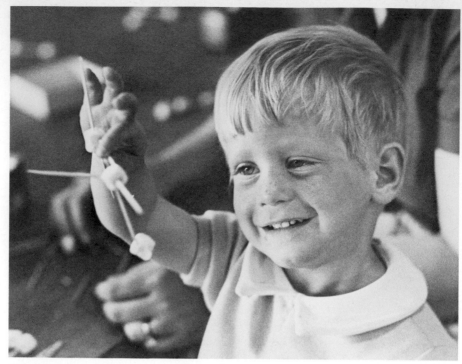

Figure 1-3 It is not easy for an adult to understand a very young child. (Photograph by Emily Richard.)

visited, science is the subject which is taught at the end of Friday *if* the art projects are finished or, worse, if the children have "been good" all week. Another facet of the problem is the way many of us have been taught science in the past. How many times have you heard the dreaded command, "Read the chapter and answer the questions at the end"? How many times did a science lesson consist of your passively sitting at your desk watching Mr. or Ms. Wizzard performing "magic" which never made any sense to you? How many times did you sink into lethargy while your encyclopedic teacher droned on and on in a monologue with the chalkboard?

The feeling of being out of touch with today's science in the elementary classroom is another source of uneasiness. This, too, is a genuine reason for concern. Science education has changed dramatically since the launching of Sputnik by the Soviet Union in 1957. That small satellite did more to agitate interest in science education in the United States than any single event, person, publication, or proclamation. All kinds of people began to wonder about our national prestige and our scientific prowess. Are our students falling behind? Are our science programs inferior? Do our science teachers need more training? "Yes" answered Congress as never before, and billions of dollars were appropriated to improve the status of science education and scientific research, especially in the 1960s and to a somewhat lesser degree in the 1970s. (We will consider in more detail some of the results of these efforts in later chapters, especially Chapter 9, Analyzing and Interpreting Elementary Science Curricula.)

Sputnik stimulated a renascence in science education from which numerous new trends arose and are still in the process of evolving. I tend to perceive the trends of science education in terms of the "big three" areas of education in general: the cognitive, psychomotor, and affective domains. (This clustering may seem contrived, and in fact I've always found it somewhat in conflict with my interest in the whole person; but I will use it now mainly for the sake of conventionality.)

In the *cognitive domain*, the following shifts are occurring:

1. From giving answers to questions children never asked to providing activities which either are based on children's questions or stimulate questioning and the search for answers
2. From requiring the memorization of meaningless terms to experiencing examples of concepts before their names are presented
3. From using spasmodic nature studies or object lessons (someone brought in a butterfly, so today we'll learn about butterflies) to treating the life, physical, and earth sciences in a more balanced way
4. From saying, "Read the text and answer the questions *correctly*" to saying, "No need to read science; just do science," to combining the reading, doing, and thinking of science
5. From creating science programs with narrow authorship to creating programs by means of broad teams including scientists, psychologists, science educators, and elementary teachers
6. From presenting science in a narrow, compartmentalized way to approaching it in a more comprehensive, multidisciplinary way incorporating mathematics, social studies, and language arts
7. From using random or tradition-bound topics (such as the seasons or night and day) to selecting topics which are conceptually more appropriate to children's understanding.

In the *psychomotor domain*, the following shifts are occurring:

1. From listening or watching or reading about science to *doing* science by means of the process skills of science
2. From defining *learning* as the momentary accumulation of facts to designing more meaningful conditions for learning and recognizing evidence of learning as an internalized long-term change in behavior
3. From advocating *the* (one) scientific method to using a variety of "modes of scientific thought"*

*In the early sixties, Robert Stollberg and a committee of the National Science Teachers Association made a very deliberate attempt to abolish the erroneous notion that all scientists adhere to one scientific method in a step-by-step fashion. Instead, they suggested the following as modes of thought generally used in any order by most scientists: (1) being bothered by inconsistencies and wanting explanations, (2) observing with discrimination, (3) classifying observations and information, (4) quantifying observations, (5) pursuing hunches and insight, (6) synthesizing and modifying explanations, (7) testing predictions, and (8) communicating results (Archie Lacey, *Guide to Science Teaching*, 1966, p. 3; reprinted by permission of Wadsworth Publishing Company, Inc.).

4. From giving one form of instruction to "one thirty-headed child" to using a great variety of activities individualized for many different children
5. From giving the excuse "We can't have science because we can't afford all that fancy equipment" to having available many inexpensive materials and needing less specialized equipment.

In the *affective domain*, the following shifts are evident:

1. From being interested mainly in what stems from scientists' heads and hands (facts and technology) to emphasizing their hearts and "what makes them tick" (values of science and technology and their affects on society)
2. From confining science to technical tomes as "science for science's sake" to pursuing scientific research that can be put into action for the betterment of humanity
3. From reassembling and redistributing a body of facts to reawakening the spirit of science as a mode of rational thought and behavior.

Don't be discouraged if you feel that you don't immediately grasp the significance of these trends. At this point, try to consider them nonjudgmentally. They will make sense only in terms of the experience you bring to them. I hope their meaning will become clearer to you as you read the other chapters, consult the Suggested Readings, and simply gain more experience; then, you can decide to what extent you agree or disagree with them.

Some Problems of Fear or Insecurity

The final reasons why many of us find science a difficult subject to teach fit under the broad category of fear of venturing into the unknown. This hesitancy might also stem from a resistance to change in general. The more uncomfortable you are about doing something, the easier it is to procrastinate or avoid the task entirely. In addition, there may be personal phobias or biases to overcome. It's not uncommon to hear remarks like these in teachers' lounges: "My children have been asking to make their own terraria." (See Figure 1-4.) "Can you imagine the mess?" Or: "My children want me to take them to the pond today. You won't find me touching that slimy gunk, much less let them bring pond creatures into my classroom." Or: "I don't mind those units on plants and animals or weather. But please don't ask me to teach about machines or electricity. The thought of it makes me break out in hives." Or: "I get flustered whenever we have to have a science lesson. That's the only time those kids ask me questions I can't answer. Why, the other day, Richard asked me to explain . . ." Or: "Just look at this new science program. Half of this stuff even I don't understand. How can I be expected to explain it to my class? How will we ever cover all that material?"

There is no denying that we are living in the midst of an explosion of knowledge—the like of which no other generation has ever experienced. Plastics, synthetics, electronic devices of all sorts, computers, lasers, and a nearly

Figure 1-4 Making a terrarium. (Photograph by Emily Richard.)

endless list of chemicals are just a few of the products that were unknown less than a generation ago. Nuclear fission began in the 1940s. Sputnik was launched in 1957. Only a dozen years later, in 1969, American astronauts were walking around on the moon's surface collecting samples, observations, and data. In 1976 the arm of the Viking planetary laboratory was scooping up Martian soil to be tested for evidence of life by the biological equipment on board. Viking contained not only a biology laboratory but also a weather station, numerous cameras, a seismograph to measure "marquakes" (or whatever one calls tremors on Mars), and of course the famous robot arm. It communicated with headquarters on Earth, 213,000,000 miles away. The space shuttle will open additional horizons. Then what? Norman Avery ponders: Will the space shuttle be our taxi service toward tomorrow? Will a space colony be our fifty-first state? (1977, pp. 45–49).

Many writers predict that the amount of scientific knowledge discovered will *double* every ten to twelve years—think about that for a moment. In the foreword to his book *Time Out for Tomorrow* (1977), Avery comments that today's infants are born into a time in which half of the "knowledge" that will be discovered before they reach the age of 20 will no longer be considered true by the time they are 40. This seems incredible until we realize what changes we have experienced in recent years.

It should come as no surprise that this explosion of knowledge, with its effects on technology, elicits fear in some and insecurity in others—and not just in elementary teachers. (It also instills great hopes in many.) Your degree of fear or lack of confidence will diminish considerably as you come to realize that science is more than content and separate from technology. Chapters 2 and 3 will give you clearer ideas on how children can deal with content. Part Two of this book, Chapters 4 through 7, should enhance your understanding of the process component of doing science with children. But all this depends on you. Again, I ask you to seriously consider the fifth premise mentioned in the Introduction: your willingness to work at becoming as good a science educator as you can. Your inadequacies, limited understanding, or fears may be real. But if you are willing to try, you can supplant them with new skills and knowledge. Is it worth all the effort? Perhaps you can decide better after considering the next question.

Why Is Science So Important for Children?

At first glance, it seems much easier to see the importance of science in society in general than in the elementary school. We can immediately see the multitude of benefits from scientists' research and technologists' applications in medicine, industry, transportation, agriculture, electronics, and so on. Think about what your day would be like if all of the scientific advances and applications that have been made in your lifetime alone were suddenly erased. How very different life would be!

But a more important matter to ponder is this: What difference does it make if science is or is not a part of your elementary school curriculum? This is a situation in which you have far greater control. You alone, obviously, cannot control the effects of science on society. But you have real decision-making opportunities in your own classroom. You can include science as part of the students' basic program; you can even emphasize it. You can also ignore it or give it only minimal "letter of the law" attention. Your judgment about its importance will influence your ultimate decision.

If you have read other books on teaching science or on elementary school curricula, you may have encountered a variety of reasons why science is said to be important for children. You might be able to recite some explanations, for example: "Science helps children become better decision makers and critical thinkers who will become more scientifically literate future citizens." These are worthy and noble goals, but there is a danger that they may become meaningless conventionalities like "Eat your spinach." How meaningful are they really? They need to be analyzed more carefully, along with some additional reasons.

Science Can Help Children Become More Rational Thinkers

What is it about science or the scientific enterprise that is conducive to more rational thinking? It would seem that a chasm exists between the thinking ability of scientists and that of children. Can scientists' modes of thinking and modes which can be acquired by children possibly have any common goals? The Educational Policies Commission presented some of the most meaningful answers to these questions in their publication *Education and the Spirit of Science*. They identified the following seven characteristics as the spirit or values of the scientific enterprise:

1. Longing to know and to understand
2. Questioning of all things
3. Searching for data and their meaning
4. Demanding verification
5. Respecting logic
6. Considering premises
7. Considering consequences (1966, p. 15)

These are goals which are applicable not only to science but to rational thinking in general. They are, therefore, relevant to the *entire* educational enterprise; by no means are they sacrosanct just to scientists. In fact, the first two (longing to know and questioning all things) are often more characteristic of children than they are of many scientists—especially scientists who have become highly specialized. For children, the entire world of their perceptions is the laboratory. Though their efforts are often fumbling, children readily search for data and want verification. Their inquiries seem endless: "What's this?" "How come?" "Why?" "When?" "Where?" (Figure 1-5). But what happens to our spirit of inquiry as we progress up the educational ladder? Why does the number of questions decrease? Perhaps part of the reason is a lack of opportunity to use scientific skills or thinking processes, so that a form of atrophy has set in.

There is no better way to help children get "back to the basics" of *their* wanting to know, *their* questioning and searching, than to allow them to interact with objects and events of the natural world. That is what's involved in their doing science, which is important to them and which should become more meaningful to you.

Figure 1-5 Children are used to asking questions of nature. (Photograph by Nancy Guyton.)

Science Can Prepare Children to Become Better Decision Makers

Because of the explosion of knowledge, more than ever before we must prepare today's children for the unknowns of the future. Impossible? Not really—since the tools or processes of rational thinking are valuable at all times and in all places. But it is highly doubtful that children will automatically acquire the thought processes involved in effective decision making. The examples of many of *our* predecessors and contemporaries are not always encouraging—particularly as regards respect for logic, consideration of one's premises (both strengths and weaknesses), and, most important, consideration of the consequences of decisions. For example, we've often failed to consider the consequences of our ill treatment of the Earth. Certainly, more rational thinking can improve future decisions, *if* we begin preparing our students now. But this is no easy task. During my adolescence, I can remember being introduced to the concepts of molecules, atoms, nuclear fissions, atomic bombs, and atomic energy. There was no warning or preparation, however, for my eventually being given the opportunity to vote on a nuclear power plant in my state. Perhaps you recall your introduction to the concepts of genes and chromosomes, and the models of DNA and RNA—and the general notion that all of these had something to do with inherited characteristics. But are you prepared to handle the decision making involved in genetic engineering, genetic screening, or cloning? It is quite possible that your students will face these decisions in the future.

We cannot provide children with all the factual information they will need in order to handle their future decisions well. As was mentioned before, much of the knowledge they will need has not yet been made available, or even discovered. If children begin to experience science as tentative knowledge and as the continuous human search for clearer answers, in which each new discovery opens the door for another, they will not be so shocked by the future. They will also be developing the basic notions of scientific literacy which are increasingly needed in society—that is, what science is about (the products) and how science operates (the processes). If they understand and have the opportunities to use the seven values recommended by the Educational Policies Commission, along with the skills to be presented in Part Two of this book, they will be well equipped to make more rational decisions both now and later. More succinctly, science is important to children because it helps them make sense out of the world.

Science Is Natural for Children

Although the explanations just discussed are very important, there is a far more fundamental reason why science is important to and for children. This most basic reason is that children have a very strong, natural affinity toward science—even the children of today's synthetic, artificially coated world.

One of the fundamental tenets of instruction is that new knowledge and skills should closely match or parallel what the learner already knows or can do. (For example, a very young child who is beginning to master the skill of walking can learn to roller skate more quickly than an older person who learned to walk years ago. The skills needed in learning to skate—such as balance and coordination—are similar to those needed for walking. Thus it is *more natural* for a young child to learn to skate than it is for an older person.) The teaching and learning of science complement the child's essential characteristics: curiosity, willingness to take risks, and a nearly inexhaustible appetite for activity. Therefore, science is "a natural" for children; it parallels or matches their basic inclinations. I've yet to meet a child who was not fascinated by "scientific stuff," who did not love to explore. Rachel Carson best portrayed the merging of science and the child's spirit, along with the role of parents or teachers, in *The Sense of Wonder:*

A child's world is fresh and new and beautiful, full of wonder and excitement. It is our misfortune that for most of us that clear-eyed vision, that true instinct for what is beautiful and awe-inspiring, is dimmed and even lost before we reach adulthood. If I had influence . . . I should ask that her gift to each child in the world be a sense of wonder so indestructible that it would last throughout life, as an unfailing antidote against the boredom and disenchantment of later years, the sterile preoccupation with things that are artificial, the alienation from the sources of our strength.

If a child is to keep alive his inborn sense of wonder . . . , he needs the companionship of at least one adult who can share it, rediscovering with him the joy, excitement and mystery of the world we live in . . .

I sincerely believe that for the child, and for the parent *or teacher* [italics added] seeking to guide him, it is not half so important to know as to feel. . . . Once the emotions have been aroused—a sense of the beautiful, the excitement of the new and the unknown, a

Figure 1-6 Preserving the sense of wonder is probably the most important reason for teaching science to children. (Photograph by Nancy Guyton.)

feeling of sympathy, pity, admiration or love—then we wish for knowledge about the object of our emotional response. Once found, it has lasting meaning. (1965, pp. 42–45)*

The need to preserve this sense of wonder is probably the most fundamental reason for teaching science to children (Figure 1-6). If you're not convinced of this, look around at the people who have lost their sense of wonder and the people who have retained it. The most important result of this sense of wonder is the quality of *caring*. We dare not allow this value to become extinct in either our children or ourselves.

Science Can Also Be Important for You

Science with children provides great potential for your own personal involvement and growth. By providing children with the kind of science experiences that will help them become more rational thinkers, better decision makers, and more scientifically literate individuals, you can ignite or rekindle your own sense of wonder and care. If you can view teaching and learning basically as

*Reprinted by permission of Harper and Row from *The Sense of Wonder* by Rachel Carson, 1965.

interpersonal relationships, you too can be caught up in the endless search—"Let's find out together." You too can discover (or rediscover) that going from the unknown to the known has its own inherent, intrinsic, self-generating, self-propelling power of excitement. *Science with Children* contains both my tentative declaratives (some ideas about science) and my strong imperatives (*do* science with children). *Your* science with children can become an ever-expanding experience of acquiring more knowledge of science and applying more of the skills of science (Figure 1-7). I hope it will become the basis of an attitude which helps *you* make more sense out of your world.

Figure 1-7 Science with children can become an ever-expanding experience for the teacher. (Photograph by Emily Richard.)

Encounters

1. State your best definition of *science*. Compare it with definitions of others, especially your colleagues. In what ways are the definitions similar and different? How could you improve your definition?
2. Ask a variety of teachers (preschool, elementary, junior high, senior high, and college) to describe how they feel about teaching science to children or teaching science in general. If possible, observe how their teaching reflects their viewpoints.
3. Describe your memories of science classes when you were an elementary student. Why were these science classes important or unimportant to you?
4. In what ways do you agree, disagree, or both with the reasons presented on (a) why the teaching of science is a problem for many elementary teachers and (b) why science is important for children?
5. Begin examining some elementary science curricular materials. In what ways do these materials present science as noun? As verb? How do they enhance or capitalize on the child's sense of wonder?
6. Specify your initial ideas of what skills and knowledge you think you need in order to do a better job of science with children. Refer to these ideas periodically, especially after you have completed this book.
7. What other questions or areas of interest related to this chapter do you wish to investigate?

References

Avery, Norman: *Time Out for Tomorrow*, T.H.A.R. Institute, Raynesford, Montana, 1977.
Carson, Rachel: *The Sense of Wonder*, Harper and Row, New York, 1965.
Educational Policies Commission: *Education and the Spirit of Science*, National Education Association of the United States, Washington, D.C., 1966.
Lacey, Archie: *Guide to Science Teaching,* Wadsworth, Belmont, Calif., 1966.

Suggested Readings

American Association for the Advancement of Science Commission on Science Education: *Preservice Science Education of Elementary School Teachers*, AAAS Miscellaneous Publication 70-5, Washington, D.C., 1970. (Guidelines, standards, and recommendations for teaching the new science programs. Provides a basic framework for teacher education.)
Bronowski, J.: *Science and Human Values*, Harper and Row, New York, 1965.(An insightful classic on the meaning of science, the values of the scientific enterprise, and the responsibilities of scientists.)

Bybee, Rodger W.: *Personalizing Science Teaching*, National Science Teachers Association, Washington, D.C., 1974. (A monograph which emphasizes the importance of interpersonal relationships in the teaching and learning of science.)

Fischer, Robert B.: *Science, Man and Society*, Saunders, Philadelphia, 1971. (The roles, ethics, limitations, and general characteristics of science are described along with the distinctions between science and technology.)

Good, Ronald G.: *Science—Children: Readings in Elementary Science Education*, Brown, Dubuque, Iowa, 1972. (A collection of articles with emphasis on the meaning of science, children's thinking, and instructional strategies.)

Hurd, Paul DeHart, and James John Gallagher: *New Directions in Elementary Science Teaching*, Wadsworth, Belmont, Calif., 1969. (An overview of the new elementary science programs which includes considerations of some of the problems, developments, and trends in elementary science education.)

Leopold, Aldo: *A Sand County Almanac,* Ballantine Books, New York, 1971. (A beautiful account of an early environmentalist's sense of wonder and of caring about the earth.)

Sund, Robert B., and Rodger W. Bybee (eds.): *Becoming a Better Elementary Science Teacher—A Reader*, Merrill, Columbus, Ohio, 1973. (The articles center on becoming a better science teacher by understanding yourself, children, the curriculum, learning styles, and teaching techniques.)

CHAPTER 2
CHILDREN AND LEARNING

People go to doctors in the hope of "being healed." Parents send their children to schools in the hope that the children will "be educated." However, doctors—even with the increased occurrence of malpractice suits—generally get greater respect and recognition than educators for doing their jobs. Why is this so? As complicated as the human body and the related healing arts are, children and the arts of teaching and learning are far more involved and perplexing. If educators and psychologists really know so much about learning, why are so many words printed or spoken about the ills and failures of schools? Why do parents complain so much that their children aren't learning? Why do so many children so dislike school? When they are asked, "What did you learn at school today?" why do they so invariably answer, "Nothing"? Surely, there is more learning occurring in our schools than most people realize. Perhaps part of the problem is that neither teachers nor their critics have a clear understanding of "learning."

What Are Some Views on Learning?

Some people equate learning with the ability to memorize or recall information. Those who have learned the most, on this view, remember the most right answers. Is learning, then, simply the movement of knowledge from the teacher to the student and back again, unchanged, to the teacher? Or must something happen in between? Learning must surely be more than memorization. For example, ask a second-grader to paraphrase or explain the pledge of allegiance, or retest your own memory a week after you took a recall-type exam. There may be times when memorization is necessary, but learning will be far more meaningful if understanding is also included.

Behaviorists—including the wide variety of followers of B. F. Skinner—relate learning to *behavioral changes* resulting from responding to given stimuli. Is this learning, or is it succumbing to manipulation? Are human beings truly free to choose, or are we governed and controlled entirely by outside factors? Are pigeons and people all that similar? The critics of Skinner believe that the active learning of an intrinsically motivated student is not only different from but also far more lasting than that of one who has been extrinsically conditioned or shaped.

Behaviorists also believe that learning is the ability to perform a specific skill or task. Therefore, learning can be viewed as *changes in capability*. The teacher's major job is to specify the desired behavior, if it hasn't already been established by a curriculum committee or curriculum developers, and then design the optimal conditions in which the learning can be achieved. This view is more widely accepted and practiced, but it also has some drawbacks. Often, the performance demonstrated or the change in behavior exhibited is a very momentary, short lived change done simply to meet some external stimulus or standard. "Real learning" should result in a more permanent or persistent change in behavior or viewpoint because of what the learner has experienced, done, and consequently *internalized*.

How do you explain your own learning? When and how did you learn what you already know? How do your experiences and interpretations compare with those of others? Which viewpoints do you espouse? There are many theories and explanations about learning—some more popular or more plausible than others. Learning is not as easily identifiable by the teacher or sometimes the student as healing is by the doctor or the patient. Teaching is an important aspect of education, but learning is its major goal. The way you go about teaching will be greatly influenced by your view of learning. One of the scholars who has influenced my views of learning, my understanding of children, and my roles as an educator is Jean Piaget. The impact of his ideas is also evident in many recent developments in science curriculum. His views on learning are worth considering in greater detail and will be examined in the remainder of this chapter and in Chapter 3.

Who Is Piaget?

It is almost impossible to understand Piaget's ideas on learning or on the development of the intellect without first understanding Piaget the person. His works—more than thirty books and over 100 journal articles—and the many books and articles about him are not light reading. Many of his ideas are unique; he often uses a highly specialized vocabulary; and the difficulties of understanding are compounded by the problems of translating from French to English. But by considering some of the highlights of his life, one is in a better position to understand his ideas.

Piaget was born August 9, 1896, in Neuchatel, Switzerland. He must have been a very bright and observant youth. His first article, dealing with his observations of an albino sparrow in one of the parks of Neuchatel, was published in a natural history magazine when he was only 11 years old. During his early teens, he became a serious scholar of mollusks, and at the age of 15 he was offered the position of curator of the mollusk collection at the Museum of Natural History in Geneva. Piaget decided to decline the offer so that he could finish high school.

During his later teens, although he continued his studies in biology, he also began to investigate philosophy. His chief interest was epistemology, the study of human knowledge, which centers on questions like "What is knowledge?" and "What does it mean to know?" He approached the study of epistemology with the diligence, vigor, thoroughness, and exactitude that he had exhibited earlier. He received his Ph.D. at the age of 21, having completed his thesis on mollusks and their evolution in and adaptation to a mountainous region of Switzerland.

The next phase of Piaget's life is (at least to me) particularly impressive. Different things happen to different people once they have acquired their final degree—a Ph.D., an Ed.D., or whatever. Some go into a slump; it might take them several months or years to become professionally productive again. Others develop a chronic case of professorial pomposity, not always curable. A few are very analytical; they appraise their accomplishments, establish new goals, and progress full speed ahead. Piaget was among this last group. However, what is unique is that he set out in an entirely new direction. Although he was skilled in both scientific and speculative thought, he chose to look for a linkage between the two in the relatively young field of psychology. He became a student again and studied psychology (especially abnormal psychology, which was very much in vogue then) for two years in Zurich and then in Paris.

When Piaget was 24, he began working at the Binet Laboratory in Paris. His job was to standardize some English reasoning tests into French versions. Ginsburg and Opper give an excellent account of the types of discoveries Piaget made and the preliminary theories he formed during the 1920s (1969, pp. 3–7).

He was intrigued more by the unusual than by the usual; he was fascinated more by children's wrong answers than by their right ones. He therefore became greatly interested in studying the reasoning processes behind errors on the tests. Gradually, he discovered that children of the same age tended to give similar answers and that there was a distinct difference in *quality* between the answers of older children and those of younger children. Previously, psychologists had tended to perceive intelligence primarily in a quantitative sense; that is, the more right answers given, the higher or greater the intelligence. (Unfortunately, today there are still many advocates of this view.)

During the same period, Piaget was also studying abnormal children at one of the hospitals of Paris. He found that the usual verbal or pencil-and-paper testing procedures were practically useless with these children because of language deficiencies. He therefore designed some problems which provided opportunities for the children to manipulate materials while responding to more subtle questions dealing with their thought processes. These problems were the precursors of what we now know as the "Piagetian tasks." Piaget became increasingly interested in children's logic, in its gradual appearance, and in the different methods of thinking which children use at various ages. But it must be realized that these were only his comparatively primitive searchings and findings. Piaget worked incessantly through over forty more years of exploration and experimentation.

During the 1930s and 1940s his studies centered on children's perceptions and their understanding of the physical attributes of matter, number, time, movement, space, and geometry. He analyzed the mental operations they used, their types of symbolic thought, and their logic. He even examined (though not deeply) their moral judgments. His attempts to unravel the complexity of the web of human development and intelligence have seemed unending.

Since the 1950s he has tended to return to and refine many of his earlier ideas on epistemology (the study of knowledge in general), on causality, and on sociology. He is probably one of the most prolific geniuses of our times. By encountering only this small fraction of his ideas, you can get a sense of his indomitable search for better understanding, and his extraordinary ability to observe, question, analyze, and synthesize.

What Are Piaget's Ideas on Learning?

Piaget's theory of how children learn is not easy to explain. Many explanations of it become either overly simplified or unduly complicated. If the adage is true that one picture is worth ten thousand words, then his theory will become much clearer if you study Figure 2-1 carefully.

Interaction Cycle   The legend for Figure 2-1 should give you your best clue to the essence of Piaget's ideas on learning: namely, *interaction*. We begin with a learner who has a unique collection of experiences. At this point, don't concern yourself with a particular child or with the more popularized notion of Piaget's ages and stages of children. One of the most powerful aspects of his model is that it can apply to any learner. To simulate the "walk through" explanation of Figure 2-1, picture *yourself* as the learner in a relative state of equilibrium or

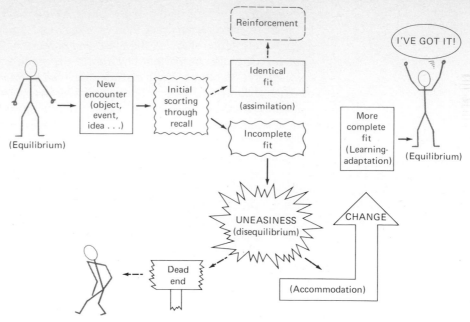

Figure 2-1   Piaget's interaction model. (From Doris A. Trojcak, "Be Careful with Those Answers," *Instructor*, November 1974; by permission of Instructor Publications, Inc.)

balance. If you say that's impossible, you probably have a good understanding of the meaning of *equilibrium*. As a scientist, Piaget chose his words very deliberately. The process of equilibration, with the resulting state of equilibrium, is often referred to in biology as well as in physics. For the most part, it is extremely short-lived. But at least we can admit that it exists and try to picture the learner in a state of *relative* equilibrium or mental rest. In order for the learning cycle to begin, the learner must encounter some sort of external disturbance or stimulus—an unfamiliar object, event, or idea. The learner then attempts to determine how well this new information fits with old information. Piaget refers to this incorporating process as *assimilation*. You, for instance, might now be sorting through your own mental filing cabinet for the folder on learning and attempting to recall how Piaget's ideas relate to what you already know.

If the new encounter matches what the learner already knows—that is, if there is a fit between previous learning and the present encounter—the learner is *reinforced*. However, note that no new learning has occurred. The slashed arrows on the diagram indicate this route. Here, that might be the case if you have already read about and understood Piaget's theory; my words would be simply supportive. Piaget does not deal with reinforcement in any great detail. The main reason why you need to consider it is that some teachers mistake it for learning. This is likely to happen with teachers who prefer to maintain the status quo, who have a "don't rock the boat" attitude. It can also happen with teachers who tend to overteach or drill unnecessarily. They fail to realize that the most important ingredients of the learning process are missing.

If the new encounter does not match what the learner already knows, the content cannot be completely assimilated (for instance, suppose that your mental file on Piaget is rather empty). Once the learner realizes this, a state of dissatisfaction or being bothered sets in. The Piagetian word for this is *disequilibrium*. It's the "I don't get it" state. Disequilibrium can take on many faces: pouting, anger, disgust, consternation, concern, confusion, curiosity, or just a mild uneasiness, to mention a few. The manifestations of disequilibrium can be overt or covert. Covert mainfestations are probably more serious and detrimental to the learning process, especially when they produce an attitude of fear or shame: "I dare not let on that I don't get this, as I must be the only one who's confused."

Here, again, the learner has a choice. One can decide to withdraw from the learning process (this is indicated on the diagram by the lower slashed arrow), ignore the new information, and simply give up. With that decision, the interaction ceases—though perhaps just temporarily. A dead end is reached. On the other hand, one can respond to the initial encounter, either physically or mentally, and thereby change or modify one's point of view or understanding as a result of the actions or mental grappling performed. Piaget calls this component of the interaction process *accommodation*.

If you are still unclear about the distinction between assimilation and accommodation, consider a common biological example: digestion. We take in or assimilate food. The body then adjusts. It goes through some changes or modifications like salivating, chewing, swallowing, contracting muscles, and releasing solutions in response to the type of food present. In other words, the body accommodates to the food. In learning, we incorporate or attempt to fit into our existing mental framework certain external stimuli or bits of reality; we assimilate. We accommodate when we change our views of behavior in response to what we have assimilated. It is the combination of these two processes that results in *adaptation*, or a more complete fit, commonly referred to as *learning*. Once again we reach the state, however tenuous, of equilibrium. This is that marvelous "I've got it" state which lasts *only* until the next accepted new encounter, at which point the interaction cycle begins again, from equilibrium to disequilibrium to equilibrium. In terms of biology, if an organism is unable to adapt to its environment—that is, take in environmental factors as well as adjust itself to them—it will not survive. In terms of intellectual development, if the learner is unable to determine how a new encounter relates to previous experiences and respond to the encounter to achieve a change in understanding, no learning will occur, and classroom or educational "survival" will not be possible.

Learners' and Teachers' Roles   Learning involves adaptation: balancing of assimilation and accommodation; responding to equilibrium and disequilibrium; and, most certainly, activity on the part of the learner. During one of his few visits to the United States, Piaget once said:

In general, learning is provoked by situations—provoked by a psychological experimenter; or by a teacher, with respect to some didactic point, or by an external situation. It is provoked, in general, as opposed to spontaneous. . . . To understand the development of knowledge, we must start with an idea which seems central to me—the idea of an *operation*. Knowledge is not a copy of reality. To know an object, to know an event, is not simply to look at it and make a mental copy, or image, of it. To know an object is to act on it. To know is to modify, to transform the object, and to understand the process of this transformation, and as a consequence to understand the way the object is constructed. An operation is thus the essence of knowledge; it is an interiorised action which modifies the object of knowledge. (1964, p. 8)

It is the *learner's* actions that constitute the essence of knowledge and, consequently, the basis of learning if we equate learning with the acquisition of knowledge. What then is the role of the teacher? If schools were to totally implement Piaget's model, would teachers no longer be needed? Hardly; but in many instances their functions and methods would be quite different. They would have to abandon one of their strongest articles of faith—the notion that teaching *causes* learning, that the learning process revolves around the teacher's actions rather than the learners'. Their greatest transformation would be from knowledge dispensers to encounter designers.

One of the most important types of decisions a teacher must make day by day, hour by hour, and sometimes even moment by moment is selecting the most appropriate items for learning encounters. If an object, event, or idea is too similar to what a learner already knows, the result may be boredom, lack of interest, or possibly reinforcement with no real learning. If the substance of an encounter is too difficult to assimilate and accommodate, the result may be frustration, anger, or failure. Consider the child who gets a steady diet of meaningless encounters. This child is gradually labeled as a "nonlearner." But many times the root of the problem is the teacher's failure to make the most appropriate encounters available. Later we will deal more directly with this most important task of providing suitable encounters for children.

Importance of Disequilibrium   We must consider the role of a Piagetian-type teacher when a child experiences the state of disequilibrium, the "I don't get it" stage. The temptation to intervene too quickly—to rescue the child from confusion, to simply give the answer and end the pain—can be irresistible. For years we've been conditioned to believe that learning should be enjoyable, that school should be fun. We have to compete with all kinds of television shows, games, and toys for children which make "learning" exciting and pleasant. Is disequilibrium all that important? It is not only important; it is absolutely vital. Piaget once said, "The goal of education is not to increase the amount of knowledge, but to create the possibilities for a child to invent and discover. When we teach too fast, we keep the child from inventing and discovering himself. . . . Teaching means creating situations where structures can be discovered; it does not mean transmitting structures which may be assimilated at nothing other than the verbal level" (1964, p. 3). Unless children are bothered, they will make no movement or will take no action to resolve the problem. Eleanor Duckworth, a former student and frequent translator of Piaget, reinforces these ideas:

As far as education is concerned, the chief outcome of this theory of intellectual development is a plea that children be allowed to do their own learning. Piaget is not saying that intellectual development proceeds at its own pace no matter what you try to do. He is saying that what schools usually try to do is ineffectual. You cannot further understanding in a child simply by talking to him. Good pedagogy must involve presenting the child with situations in which he himself experiments, in the broadest sense of that term—trying things out to see what happens, manipulating things, manipulating symbols, posing questions and seeking his own answers, reconciling what he finds one time with what he finds at another, comparing his findings with those of other children. (1964, p. 2)

Try to relate these ideas to your own learning experiences or to the experience of living in general. As long as you are satisfied with your present capabilities or status, there is no need to change. It is only when you discover that your present ways of operating are insufficient or inadequate that you will see the need to decide whether to try out new ways or to maintain your status quo. You always have this choice; disequilibrium in itself will not cause you to change. In fact, unless you recognize and accept it, there is no need to change. Children also must have the opportunity to make this realization freely according to their own level of understanding—a level which will most likely be quite different from yours but is equally important to their learning.

Potential for False Accommodation One other component of Piaget's interaction model must still be scrutinized: the process of accommodating. Jonas Langer has suggested that more emphasis has been placed on assimilation (the fitting in of knowledge) than on accommodation, or adapting to what is encountered and reorganizing one's view (1964, p. 79). Maybe we are simply more comfortable with or aware of the process of assimilation than we are of accommodation. In biological terms, it seems that most people tend to be more aware of taking food into their bodies than they are of how their bodies are acting upon that food (unless something goes wrong, of course). In terms of education, could it also be that we are more concerned with ingestion than with digestion? Could learning ever occur without accommodation, or with the wrong kind of accommodation? While attending the 1964 Conference on Cognitive Studies and Curriculum Development at Cornell University, Piaget was asked a related question. He responded: "This is a big danger of school—false accommodation which satisfies a child because it agrees with a verbal formula he has been given. This is a false equilibrium which satisfies a child by accommodating to words—to authority and not to objects as they present themselves to him" (1964, p. 4). False accommodation is a real possibility in the classroom as well as in life as a whole. Children (and adults) can change or adjust their responses, reactions, or behaviors as a result of pressures or influences from peers, parents, teachers, or "the system" rather than as a result of their own acting upon objects, events, or ideas encountered. I've seen numerous examples of false accommodation, some serious and some humorous, from both students and teachers:

A child parroting meaningless words or answers. One 5-year-old has heard the word *gravity* on television, and so the cause of *every* event is simple for him: "It's all gravity."

Third-graders announcing to their teacher, who is about to begin some activities with wire, batteries, and bulbs: "We already learned *all* about electricity in second grade."

A fourth-grader who completes a science assignment to define a list of new words not by reading the science text first but by copying the shortest definitions in the dictionary. (She receives a superior grade for her inferior "learning" and is a bit upset when I ask her to paraphrase some of the words. She feels that she has already demonstrated her "learning.")

A college student who will only reply, "I did enough to get by" when asked what he has learned from a prerequisite course. When pressed—"But did you really learn . . . ?"—he resorts to, "Well, I guess so; I passed the course."

A student teacher who gives herself an A+ on her self-evaluation report, which asked for evidence of the children's learning. She gives the following account: "*I* very clearly *explained* that the child must always be between the sun and his shadow, since it is because the body stops the sunlight that shadows are formed. These first-graders were very attentive listeners."

An experienced teacher in a graduate course who, when asked to comment on successful science teaching efforts, describes teaching the entire human body to fifth-graders: "I've got it down to twenty-nine lessons now."

Surely you could add your own examples to this list of false accommodations. Recognizing and sometimes even unmasking the false learner is a never-ending task. Unfortunately, students can become very adept at bluffing or "psyching out the teacher." Degrees of security or insecurity, trust or distrust, freedom or constriction have a great impact on learning. And regardless of our best efforts or the best conditions, it is still easy to misinterpret or be unaware of one's own behavior. Once, during an in-service session on Piaget, I noticed a woman who often nodded her head up and down, not dozing, but in apparent agreement with what I was saying. During the coffee break, she confided to me: "I want to thank you for helping me discover something about myself. I didn't realize it until today—but whenever I'm confused, I tend to nod my head up and down." Few people can ever admit so openly or recognize so accurately the presence of false accommodation. But recognition of false accommodation can be startling when it does occur. So can the recognition of disequilibrium, which is described in this final example. A new group of students were introducing themselves at the beginning of a semester. Several mentioned their involvement in various cybernetic activities. One of the participants finally exclaimed in exasperation: "I've raised three kids and now want to become a teacher. I understand A.M., P.M., and B.M., but what's this business of alpha, T.A., and T.M.?" Often children cannot be that forthright in letting you know where they stand. But we can gain much information and insight on understanding children from the research findings of Jean Piaget, which are discussed in the following section.

What Are Some Characteristics of Children?

Even though his message has not been universally accepted in practice, Piaget has probably done more than anyone else to promulgate data indicating that children are *not* miniature adults. Their thinking is distinctly different from that of most adults and is also different at various stages of their development. As a result of over forty years of research—his own and that of his colleagues—Piaget has generated an amazingly thorough model of intellectual development. These are the now popularized findings presented in just about every psychology textbook: (1) that children progress through various stages of intellectual development, (2) that the stages occur in a definite sequence, (3) that there are some general age spans associated with these stages but that children can and often do move through the stages at different rates, and (4) that intellectual development does not occur in a lockstep fashion—that is, a child might be at one stage as regards X content and at a different stage as regards Y content.

Were you to research the literature, you would most likely find a variety of names for the stages and even the variances among the age spans. But the following tend to be the most acceptable:

1. *Sensory-motor stage*—from birth to approximately 18 to 30 months
2. *Preoperational* or *representational stage*—from 18 months to 7 to 8 years
3. *Concrete-operational stage*—from 7 to 8 years to 11 to 12 years
4. *Formal-operational* or *propositional stage*—from 11 to 15 years and onward

Again, remember that the *sequence* of the stages is constant, but the ages are *not*. For example, it would not be unusual for a fifth-grade teacher to face a class including preoperational, concrete-operational, and perhaps even formal-operational thinkers. Since you will have the opportunity to relate to a variety of children during your teaching career, it is important to understand the characteristics of children in each of the stages of development. It is also important to put Piaget's efforts and findings in the proper perspective. Too often his results appear as a "grocery list" of traits—the sort of finished product that can readily be placed in the "memorize these for the test" category. It is extremely difficult to capture the progression of his own reasoning processes or the interpretation of them by others. His ideas evolved gradually and only after hundreds of hours of observing, analyzing, and synthesizing the behaviors of children. Perhaps if you can sense the process of the development of ideas (both Piaget's and children's), the products or the actual descriptions of the characteristics will be more meaningful. It is misleading to view any of the stages as static. Try to identify the subtle but strong undercurrent of dynamic change, growth, and development. This is especially necessary for understanding the first stage.

Sensory-Motor Child (birth to 2¹/₂ years)

The intellectual growth of the child in the sensory-motor stage is limited, relatively speaking, but immensely important. Imagine spending a day with a newborn infant, departing the scene, and then returning again to spend another day with the child after eighteen to thirty months. "Is this the same infant?" you might wonder.

The newborn infant doesn't begin with much; a puppy not yet dry behind the ears has far better instincts. But as Hans Furth has noted, the infant carries a *capacity to develop* by acting on its environment far more potent than the puppy's (1970, p. 23). The sensory-motor child learns by interacting with the immediate environment, using the senses and muscles. This child uses an extraordinary variety of actions—watching, testing, feeling, smelling, listening, reaching, squeezing, hitting, squirming. Gradually, the seemingly purposeless antics become more purposeful explorations. But throughout this stage, the child's behavior is basically reflexive, not reflective. Knowing is based entirely on sensory-motor doing, not on representational language—that is, not on using words to represent what is known. The rudiments of language begin to be acquired toward the end of this stage. But one must have a special knack to determine to which object, event, or idea the child is attaching a sound. According to Piaget, the sensory-motor child sees life as if viewing a slow-motion film. The individual pictures appear in succession, but they are not integrated and therefore form no meaningful composite.

Yet it is during the sensory-motor interaction period that the child is experiencing and collecting the material which will form the foundations of later learning. By approximately 18 months, the child will have developed the concept of object permanency. Prior to this discovery, if an object is out of the child's perceptual field, it simply does not exist. ("Out of sight; out of mind.") The child develops other perceptual skills such as the sense of space, temporal succession (different events normally occur one after the other), and sensory-motor causality (an open flame equals "ouch"). The importance of the sensory-motor doing of the very young child cannot be overemphasized. The child can enter the next stage either greatly enriched or already impoverished, depending on the experiences he or she has had of acting upon the environment.

Preoperational Child (18 months to 7 to 8 years)

During this stage, which is also called the *representational* stage, the child exhibits the greatest growth in language development. Gradually words come to be used by the child to represent thoughts (in the sensory-motor stage, only actions are used). At first these verbal labels are related only to immediate objects, events, or wishes; later they are used to refer to objects, events, or mental images which are not present. However, these absent entities must be rather immediate, since the development of memory is still limited. By listening to and later imitating the language of adults and older children, the peroperational child gains an ever-expanding view of the environment. This is an extremely exciting time for the child but a potentially risky period for the adults in the children's life. Many adults fail to realize the very limited nature of the child's understanding of the concepts and of the words used to represent these concepts. We are tempted to believe that even at this very early age language acquisition and reasoning ability are synonomous. Ginsburg and Opper state: "Language does not completely impose on the child the culturally desirable ways of thinking. Instead, the child distorts the language to fit his own mental structure. The child achieves mature thought only after a long process of development in which the role of language is but one contributing factor" (1969, p. 85).

The key to understanding some of the other contributing factors is contained in the more frequently used descriptor of this stage: *preoperational*. The prefix *pre* is simple enough: "coming before" or "preceding." But what does *operational* mean in the Piagetian sense? Or, what is his concept of an operation? Piaget considers an operation to be the very essence of knowledge. An *operation* consists of interiorized actions used by a learner to cope with an object of knowledge. Operations are mental actions, such as classifying a collection of objects, putting things or events in a serial order, counting, and measuring. An operation also implies a *reversible* action; that is, the mental activity can go in either direction, such as adding or subtracting, joining or separating, comparing one set of variables against another set (Piaget, 1964, pp. 8–9).

Why is the child between the ages of 18 months and 7 to 8 years generally unable to demonstrate these kinds of operations or mental actions? The major reason—and perhaps the basic reason for the vast differences between young children and adults or even older children—is that the peroperational child is *perception-prone* or perceptually predisposed. This child judges by perceptions—by what is seen—rather than by logical reasoning. If something looks different, it *is* different. Chittenden has aptly described the child's reasoning as a logic of convenience rather than a logic of conviction. Contradictory explanations do not disturb the preoperational child. The preoperational child does think—but not about his or her own thinking (1970, p. 11). An example of one of the classic Piagetian tasks might help. A child is given a ball of clay to handle and then asked to roll the clay into a "snake" or sausage shape. The interviewer next asks something like, "Do you now have more clay in the snake? Or less? Or do you have the same amount of clay that was in the ball?" The preoperational, perception-prone child will most likely respond in one of two ways: "There's more clay because look how long the snake is." Or, "There's less clay because the snake is so skinny." These are typical responses from a *nonconserver*. *Conservation*, another Piagetian concept, relates to the ability to understand that the original amount of something remains the same regardless of how the physical appearance has been altered—provided, of course, that something has not been added or taken away. Since the clay snake looks so very different from the clay ball, even the amount of clay has become different to the preoperational child.

The responses to the clay-ball task also indicate another characteristic of a perceptually predisposed child: the tendency to *center*. The preoperational child tends to attend to just one attribute of an object or event and ignore all other characteristics. The combined thoughts of length and skinniness, of color, texture, mass, and odor would not spontaneously occur to the preoperational child. Neither would the notion of *reversibility*. That is, the child would not consider reversing the clay (either mentally or physically squeezing it) back to the original spherical shape. As was stated previously, an operation implies a reversible action. This is beyond the capability of the preoperational child, whose thinking is irreversible or one-directional.

This child is also unable to cope with states of *transformation* or changes in a sequence of events. Attention is given mainly to the beginning and ending of a series of changes. The in-between events are either confused or ignored. (Two examples of transformation activities will be explained in Chapter 3.) This inability to recognize states of transformation, or the tendency for thought to be static, is related to centering and irreversibility.

Another characteristic that is pronounced in preoperational children, but certainly not unique to them, is *egocentrism*. The egocentric child is the center of the universe about which all else revolves. There is but one point of view—his or hers. Since the preoperational child is incapable of considering more than one variable at a time, it should not be too surprising that the egocentric child is unable to consider another person's viewpoint simultaneously with his or her own. (What is far more surprising and frightening is the number of adults who cannot do this.) A child (as well as an adult) is constantly faced with the need to decenter from one's own narrow viewpoint. There is a strong inverse relationship between egocentrism and learning. As the former decreases, the latter increases.

An inverse relationship probably also exists between learning and the last characteristic of the preoperational child that we will consider—*transductive thought*. You've experienced inductive reasoning, going from specific to general; and the opposite, deductive reasoning, going from general to specific. But you might have forgotten or been unaware of the third type. In transductive reasoning one goes from specific to specific. Renner, Stafford, and Ragan (1973, p. 77) describe a typical transductive response in one of their child studies. A 4-year-old was shown many wooden beads, mostly red ones and some blue. The child was asked if there were more wooden beads or red beads and responded in typical preoperational fashion that there were more red beads. When asked why, the child responded, "Because they are prettier." Adults also sometimes use transductive reasoning. The mother of a friend of mine had made an illegal left turn and was stopped by a police officer. When the woman was asked if she realized that she had made an illegal left turn, she said, "Well, of course; it's Wednesday." It was, in fact, Saturday; the policeman was so befuddled that he simply watched her drive off. Surely you've experienced other examples of transductive reasoning, and not just in young children.

Remember that the ages noted for Piaget's stages are only approximations for the majority of children. You may know some young children who do not fit the characteristics of the preoperational child. You may also know some adolescents and adults for whom many of the characteristics of young children are amazingly suitable. (I once visited a developing country and returned with the feeling that I had experienced a preoperational nation.) You must also realize that intellectual development is continuous and gradual. A person does not simply leave one stage and enter the next as though walking through a turnstile. It is highly possible that one may have some sensory-motor as well as some preoperational traits. According to Hans Furth, the child on the path of preoperational thinking adapts laboriously to the use of symbols and slowly reconstructs on the operational level that which was already achieved on the level of sensory-motor coordination (1969, p. 67). Let's next examine the characteristics of the child on the road of concrete-operational thinking.

Concrete-Operational Child (7 to 8 years to 11 to 12 years)

The better you understand the preoperational child, the easier it will be for you to understand the concrete-operational child. Unlike the preoperational young-ster, this "oldster" can perform mental operations or the basic logical operations. However, this child can perform these mental operations *only* on concrete or real objects, *not* on abstract ideas or verbal generalizations or even on the printed word. (Those modes of instruction or learning are for the next stage.) Because the concrete-operational child appears to be so much "smarter" than the illogical preoperational child, parents and teachers sometimes become fidgety, jump the gun, and expect the concrete-operational child to function on a more abstract or adult level. Fifth- or sixth-grade teachers have sometimes been guilty of lecturing to their classes, expecting the students to take copious notes, urging them to "study your notes," and then wondering why there is so little evidence of the students' mental operations on the pencil-and-paper tests. Learning for the concrete-operational child stems from interacting with objects, not with symbols, ideas, or abstractions. The ability to mentally operate on objects by seriating, classifying, using numbers, measuring, recognizing spatial and temporal relationships, inferring, and predicting are the forte of the concrete-operational child. These skills were only embryonic in the preoper-ational child.

The most obvious difference between the preoperational child and the concrete-operational child is the latter's ability to conserve. The concept "conservation" has probably been researched more than any of Piaget's other ideas. Six different types of conservation have received the most attention: number, substance, length, area, weight, and volume. Again, the notion of sequence and ages occurs. However, unlike Piaget's four stages of intellectual development, the sequence in which the types of conservation are acquired does vary somewhat, though not significantly. The ages at which children develop the different types of conservation tend to vary more than the sequence. Dyrli gives the following definitions and examples:

1. *Conservation of number (6–7 years of age):* The number of elements in a group remains the same regardless of how the group is arranged in space. Example: The number of marbles spread out in a chalk tray does not change when the marbles are placed together in a cup.
2. *Conservation of substance (7–8 years):* The amount of a substance remains the same regardless of how its shape is altered. Example: If a container of paint is used to fill several jars, the total amount of paint does not change.
3. *Conservation of length (7–8 years):* The length of an object or a line remains

unchanged (discounting expansion or contraction) no matter how it is displaced in space. Example: A straight wire bent into a circle remains the same length.

4. *Conservation of area (8–9 years):* The total amount of surface covered by plane geometric figures remains the same no matter how the figures are rearranged. Example: The amount of area covered by ten dominoes remains the same whether they are close together or far apart.

5. *Conservation of weight (9–10 years):* The weight of an object remains the same no matter how its shape is altered. Example: A bag of potato chips weighs the same even though the chips are pulverized.

6. *Conservation of volume (14–15 years):* The amount of liquid that an object displaces remains the same regardless of how its shape is changed. Example: A ball of clay that causes the level of water in a beaker to rise 10 ml [milliliters] when submerged will cause the same volume displacement even when fashioned into a pancake-like shape. (1970, pp. 126–127)*

By the end of the concrete-operational stage, the child will have acquired most or all of the conservation concepts except volume. (A high percentage of adults are still nonconservers of volume.)

Why are these conservation concepts generally attainable by the concrete-operational child and not the preoperational child? Apparently the most basic reason is that the concrete-operational child is no longer bound to judge only perceptually as the preoperational child is. This freedom enables the child to *decenter.* For example, when the concrete-operational child rolls a ball of clay into the form of a snake, a variety of different mental operations or reasoning processes are evidenced. The clay snake is not just longer *or* skinnier; it is both longer *and* skinnier. In other words, the child notes a reciprocal relationship or *compensation* among the attributes: the short, fat ball of clay is equal to the long, skinny snake. Or the child might deduce a *logical necessity* such as, "You didn't add any or take any away; so it's got to be the same." The child can also perform *reversals* in thought, action, or both and reshape the snake into the original sphere. Decentration also occurs when the child notes other changes in forms and events. The sequential relationships among various states of transformation can be detected—not just the beginning event and end result, which are all that can be recognized by the preoperational child.

The concrete-operational child is less egocentric, better able to coordinate others' viewpoints, and consequently more objective. Some rather simple attempts at inductive and deductive reasoning may begin to appear, but the residue of egocentrism and transductive "logic" still lingers.

The concrete-operational child has developed into a many-splendored young thinker. It is a gross oversimplification to identify this child simply as "a conserver" (or the preoperational child as "a nonconserver"). This child is a composite of many characteristics and thought processes, all interrelated. The dispersed activities of the sensory-motor child and the preoperational child are

---

*Reprinted by permission of Prentice-Hall, Inc., from Odvard Egil Dyrli, *Developing Children's Thinking through Science* (James Weigand, ed.), 1970.

finally converging toward more logical thought. The fruition of these interrelationships can be seen in the fourth and final stage.

Formal-Operational Child or Adolescent (11 to 15 years and onward)

A brief review along with an initial orientation might be useful at this point. We've already considered how the preoperational child was freed from the more or less random, reflexive behaviors typical of the sensory-motor child, and how the concrete-operational child was liberated from reliance on perceptual judgments alone. We're now ready to examine how the formal-operational child or adolescent is freed from being dependent on operating only on objects in order to learn. You should now be able to recognize that (1) sensory-motor children don't really reason; they simply act and react; (2) preoperational children do very little (if any) thinking about their thinking; and (3) concrete-operational children can think if they have objects on which they can perform their mental operations.

Now we come to the end product of Piaget's developmental stages: the formal-operational child or adolescent (or adult) who can think, who can think about thinking, and who can think about the implications of the thinking. He or she is able to reason with symbols, ideas, abstractions, and generalizations, and not only on immediate objects or events. This person is able to go beyond the tangible here and now. How is this accomplished? What unique qualities or special skills do formal thinkers have that the others lack? In the literature, explanations such as "hypothetical-deductive operations," "combinational systems," and "propositional logic" appear. This terminology is typical of discussions of this stage. As we examine it, you may encounter not only some unfamiliar concepts but also some specialized vocabulary.

One of the distinctive characteristics of formal thinkers is evident in their approaches to problem solving. Their methods of acting upon encounters tend to be much more organized, integrated, and systematized. For example, when a formal thinker interacts with a problem to be solved, the following "steps" tend to occur:

1. Speculation about various possibilities—what could occur or what might be feasible
2. Observation of data and realization that what has occurred is but one possibility
3. Formulation of a hypothesis
4. Conduction of controlled experiments in which all factors are held constant except the one variable being tested
5. Deduction of conclusions
6. Confirmation or refutation of hypothesis

These *hypothetical-deductive operations* may appear vaguely familiar. Aren't these steps basically the same as those memorized as the "scientific method"? In some older elementary science programs, the "scientific method" was presented as early as grade 3 or grade 4. The children were urged to use the steps (in lockstep fashion) in order to develop the habit of thinking scientifically. But the preoperational or concrete-operational child does not have the mental operations to handle hypothetical- deductive operations. For the formal thinker, these operations are personal, internalized actions rather than impersonal tactics imposed from the outside. They can be used habitually by the formal-operational child, adolescent, or adult. If you wish to examine these thinking operations in greater depth, Ginsburg and Opper give a thorough explanation (1969, pp. 181–206).

The formal thinker's problem-solving ability is also enhanced by the use of additional operations of logic. Any of the following four might be selected:

1. *Conjunction:* The results are A and B. Example: His temperature is 38° C, and he is feverish.
2. *Disjunction:* The result is either A or B. Example: It's a cold, damp, overcast day; and so it will either sleet or snow.
3. *Negation:* The result is neither A nor B. Example: Neither the amount of light nor the amount of moisture affected the growth of that microbe.
4. *Implication:* If A is true, then B will occur. Example: If the liquid is heated, the powder will dissolve faster.

The formal thinker can use any of these types of *propositional logic* during hypothetical-deductive operations. They can also be used simply to derive inferences, predictions, hypotheses, or conclusions in general.

Such high-powered thought processes are possible for the formal-operational person because of the acquisition of *combinational systems*. That is, one can now consider all possible combinations to a problem rather than test each factor in isolation and haphazardly as the concrete-operational child does.

There are all kinds of instances in your *everyday* activities where you might need to make systematic combinations. For example, if you were planning a trip, you might need to consider whether to (1) be away a few days or many days, (2) take a large or small suitcase, (3) use cash or travelers' checks, and (4) travel by plane or car. How many combinational choices would you have? Some people might be too exhausted to even take the proposed trip after finally figuring out the sixteen possible combinations. Others, who have developed the ability to use combinational systems, would be ready to go in no time.

The formal-operational thinker has the most sophisticated set of mental tools. He or she has the ability to work on both the possible and the ideal as well as the skill to distinguish between them. Hypothetical-deductive operations, propositional logic, and combinational systems are the tools of logical reasoning. These operations also enable the propositional thinker to speculate and imagine the widest variety of possibilities. Compared with the sensory-motor child (who has no reasoning ability), the preoperational child (who is relatively illogical), and the concrete-operational child (whose operations are advanced but nonetheless limited), the formal-operational thinker has virtually unlimited capacity.

Additional Considerations

You've now had an opportunity to analyze the characteristics of children and how they learn, at least according to Piaget's views. You should have some tentative answers to the questions in this chapter. But I imagine that you are still experiencing a certain degree of disequilibrium. At this point you might be wondering: What if *I'm* not a formal operational thinker yet? Will I ever be? Just what are the implications of all this? It's interesting information, but what good will it do me as an educator? How do I apply this? Does Piaget have all the answers? I hope you will gain more insight toward resolving these questions in Chapter 3.

Encounters

1. Identify several specific areas of content or skills which you feel you learned very thoroughly at any point in your life. Analyze these learning experiences in terms of the following questions:
    a. How did you learn whatever you listed?
    b. How does this learning compare with other learning which was less thorough or less meaningful?
    c. When did you realize that the learning had occurred?
    d. How did your learning experiences compare with Piaget's ideas on the way children learn?
    Compare your findings with those of others. What are the most significant similarities?
2. Describe several examples of false accommodation which you have demonstrated yourself or witnessed in others. What do you think were the reasons or causes for each of the examples? (Try to be as specific as you can.)
3. Make a list of what you consider to be the most important characteristics of children in each of the four different stages of intellectual development. Which characteristics do you think hinder learning the most? Which characteristics aid most in intellectual growth or learning?
4   Spend some time observing science encounters in preschool, primary, intermediate, junior high, senior high, or college classes. Describe and analyze the suitability of the encounters. Which resulted in reinforcement, adaption-equilibrium, or a dead end? What changes would you recommend for the instances that resulted in a dead end?
5. In what ways do you agree with Piaget's interaction model? In what ways do you disagree with it?
6. What do you think are the main implications for education of Piaget's theories? Describe the ideal Piagetian classroom. How would the classroom look? What types of activities might you see children doing? What might the teacher be doing? How would this classroom setting compare with what you experienced as an elementary student?

7. What other questions or areas of interest related to this chapter do you wish to investigate?

References

Chittenden, Edward A.: "Piaget and Elementary Science," *Science and Children*, December 1970.

Duckworth, Eleanor: "Piaget Rediscovered," in *Piaget Rediscovered*, Report of the Conference on Cognitive Studies and Curriculum Development, National Science Foundation and United States Office of Education, March 1964.

Dyrli, Odvard Egil: "The Learning of Science," in James Weigand (ed.), *Developing Children's Thinking through Science*, Prentice-Hall, Englewood Cliffs, N.J., 1970.

Ginsburg, Herbert, and Sylvia Opper: *Piaget's Theory of Intellectual Development: An Introduction*, Prentice-Hall, Englewood Cliffs, N.J., 1969.

Furth, Hans G.: *Piaget and Knowledge*, Prentice-Hall, Englewood Cliffs, N. J., 1969.

————: *Piaget for Teachers*, Prentice-Hall, Englewood Cliffs, N. J., 1970.

Langer, Jonas: "Implications of Piaget's Talks for Curriculum," in *Piaget Rediscovered*, Report of the Conference on Cognitive Studies and Curriculum Development, National Science Foundation and United States Office of Education, March 1964.

Piaget, Jean: "Development and Learning" in *Piaget Rediscovered*, Report of the Conference on Cognitive Studies and Curriculum Development, National Science Foundation and United States Office of Education, March 1964.

Renner, John W., Don G. Stafford, and William B. Ragan: "The Child," in *Teaching Science in the Elementary School*, Harper and Row, New York, 1973.

Trojcak, Doris A.: "Be Careful with Those Answers," *Instructor*, November 1974.

Suggested Readings

Anastasiou, Clifford J.: *Teachers, Children, and Things: Materials-Centered Science,* Holt, Rinehart and Winston of Canada, Toronto, 1971. (Limited background information but a multitude of ideas for activities; good preparation for understanding the implications of Piaget's ideas.)

Brearley, Molly, and Elizabeth Hitchfield: *A Guide to Reading Piaget*, Schocken, New York, 1969. (Background information on essential Piagetian concepts.)

Charles, C. M.: *Teacher's Petit Piaget*, Fearon, Belmont, Calif., 1974. (An extremely brief overview of Piaget's ideas which can help you summarize your ideas.)

Holt, John: *How Children Fail*, Pitman, Dell, New York, 1970. (Excellent examples on real and false accommodation.)

————: *Escape from Childhood*, Ballantine Books, New York, 1975. (A provocative book on the needs and rights of children.)

Isaacs, Nathan: *A Brief Introduction to Piaget*, Agathon Press, New York, 1972. (Good background on conservation concepts, especially number.)

Maier, Henry W.: *Three Theories of Child Development*, Harper and Row, New York, 1965. (Studies and applications of Erikson's, Piaget's, and Sears's works on emotional, cognitive, and stimulus-response behavior patterns.)

Piaget, Jean: *Structuralism*, Basic Books, New York, 1970. (A thorough, didactic explanation of logical structures in terms of wholeness, transformation, and self-regulation.)

————: *The Child and Reality*, Grossman, New York, 1973. (A readable map of the mental development of children and their understanding their world.)

————: *To Understand Is to Invent*, Grossman, Viking Press, New York, 1973. (Some of Piaget's most penetrating comments on intellectual and ethical education.)

Schwebel, Milton, and Jane Raph: *Piaget in the Classroom*, Basic Books, New York, 1973. (Actual studies of classroom practices and results from employing Piaget's theories.)

Sime, Mary: *A Child's Eye View*, Harper and Row, New York, 1973. (A succinct but thorough overview of the preoperational, concrete, and formal stages.)

CHAPTER 3
APPLYING PIAGET'S IDEAS

WHAT ARE SOME IMPEDIMENTS TO APPLYING PIAGET'S IDEAS?
Parents
Administrators
Curriculum
Disequilibrium

Understanding Piaget's ideas is far more difficult than memorizing strange terminology or a list of characteristics of children at different levels of intellectual development. The best way to comprehend Piaget's ideas about children is to have face-to-face encounters with children. You need to see how children learn by becoming a part of their learning encounters. Then you can experience and examine your own adaptive processes. What ideas will you incorporate? What changes will occur in your own viewpoint? You must gather your own evidence to either support or refute Piaget's ideas on what children are actually like and how they really learn. Only then can you begin to realize the implications of his theory and choose to implement those which you value. Your applications of Piaget's ideas will then be far more powerful and effective, not because you've read or heard something here or elsewhere, but because of what you have assimilated and accommodated as a result of your experiences.

How Can Piaget's Ideas Be Examined More Thoroughly?

There is probably a near-infinite variety of ways to approach the task of examining and analyzing how children learn and what kinds of thinking processes they use. Piaget tended to use three very basic techniques: (1) he observed and recorded children's behavior in great detail; (2) he asked children a wide variety of questions; and (3) he sometimes gave children simple materials to manipulate and posed questions related to the perceived changes in the materials used. He also listened intently and analytically to each child's responses. You can use these same techniques to examine children's thinking processes and to see how your observations support or refute Piaget's findings. The following sections provide suggestions for utilizing the basic techniques or models: observing, informal questioning, and formal interviewing (administering Piagetian-type tasks). Explanations will also be given for interpreting the results of informal and formal interviews with children. Some of the reasons why children differ intellectually will be discussed, along with some of the implications of and impediments to applying Piaget's ideas.

Observing the Characteristics of Children

Suppose you spent a considerable amount of time observing children's behavior in a variety of classroom settings. How similar or dissimilar to Piaget's

43

Table 3-1
OBSERVATIONS OF CHARACTERISTICS OF CHILDREN

	CHARACTERISTICS OF STAGES OF INTELLECTUAL DEVELOPMENT	CHILD	FREQUENCY
I. SENSORY-MOTOR STAGE (BIRTH TO 18-30 MONTHS)	I.1. Interacts with environment using senses and muscles		
	I.2. Lacks meaningful language initially		
	I.3. Begins to develop object permanency		
	I.4. Begins to develop sense of space		
	I.5. Begins to develop temporal succession		
	I.6. Begins to develop sensory-motor causality		
	I.7. Begins to develop primitive language		
II. PREOPERATIONAL STAGE (18 MONTHS TO 7-8 YEARS)	II.8. Displays greatest language growth		
	II.9. Cannot perform mental actions or operations		
	II.10. Is perception-prone		
	II.11. Centers on only one characteristic at a time		
	II.12. Is egocentric		
	II.13. Is a nonconserver		
	II.14. Uses irreversible thinking or one-directional actions		
	II.15. Cannot cope with states of transformation		
	II.16. Uses transductive thought		
III. CONCRETE-OPERATIONAL STAGE (7-8 YEARS TO 11-12 YEARS)	III.17. Uses basic logical operations on real objects		
	III.18. Is a conserver of number		
	III.19. Is a conserver of substance		
	III.20. Is a conserver of length		
	III.21. Is a conserver of area		
	III.22. Is a conserver of weight		
	III.23. Centers on more than one characteristic at a time (decenters)		
	III.24. Uses reversible thinking or actions		
	III.25. Relates states of transformation		
	III.26. Is able to consider other points of view (is less egocentric)		
	III.27. Uses less or no transductive logic		

Table 3-1, continued

	CHARACTERISTICS OF STAGES OF INTELLECTUAL DEVELOPMENT	CHILD	FREQUENCY
	IV.28. Acts on abstractions		
	IV.29. Uses hypothetical-deductive operations (speculates systematically)		
IV. FORMAL STAGE (11–15 YEARS AND ONWARD)	IV.30. Uses combinational systems		
	IV.31. Uses propositional logic of conjunction		
	IV.32. Uses propositional logic of disjunction		
	IV.33. Uses propositional logic of negation		
	IV.34. Uses propositional logic of implication		
	IV.35. Distinguishes reality from possibility		

descriptions might these children appear? How might you organize your observations? One method of recording observations is presented in Table 3-1; you can modify it in any manner that will suit your individual needs or interests.

The record of observations shown in Table 3-1 can be used for one, several, or many children over any length of time. With more practice at observing and recording children's characteristics, your understanding of Piaget's descriptors should improve. You should also become more able to judge the appropriateness or inappropriateness, the accuracy or inaccuracy of his stages of intellectual development. But if you want to become more directly involved in examining children's thinking processes, the next two models will be more useful.

Questioning: Informal Conversations with Children

In its simplest form, this procedure amounts to asking children questions which they themselves tend to ask. These might be questions like the following: Where does night come from? How do clouds move? Who makes rain? What are stars? Where does the sun go at night? How do birds fly? What (or who) makes the television work?

Do not be deceived by the simplicity of this technique. Clinical psychologists also use these procedures and need years of experience with clients before they become proficient. However, clinical psychologists tend to be interested in analyzing and interpreting behaviors, in answering the questions *how* and *why* with regard to the subjects' responses. At this point, we'll be concentrating mainly on observing what children say in response to questions about their understanding of the world.

Preliminary Considerations It is not always easy to carry on an informal interview with children between the ages of 4 and 12, or even children older than 12. There are several preliminary cautions worth considering. The first tendency which must be overcome is talking too much (unless you are one of those extremely rare persons who prefer listening to talking). As you plan your questions, you'll need to maintain a very delicate balance between spontaneity and purposefulness. In other words, you'll need to be quite natural, relaxed, free-flowing, able to interact with the child; but you must know at all times (as much as possible) exactly what you are trying to accomplish with your questions. This is an informal interview, and it is meant to be much more than trivial chit-chat with a child. Even a conversation with a 5-year-old can be serious when notions of reality and causality are being discussed. It requires real skill and sensitivity to gradually create an atmosphere in which children are willing to talk freely about their ideas, views, interests, reactions, or inclinations.

If the child you are interviewing is a stranger, it would be beneficial to spend a few moments on introductory types of questions and comments. You might ask the child to describe likes or dislikes, foods, pets, toys, games, sports, places of interest, family, or friends—anything to initiate conversation from the child. Try to avoid asking questions that can be answered simply by "yes" or "no." Chances are that's all you'll get, and you will immediately have to come up with another question. The purpose of this preliminary conversation is to find clues on how to word your more important questions. You must try to gear your vocabulary to the child's level. But despite your most conscientious efforts, the child's understanding of your words is most likely not the same as your understanding.

Once you are ready to begin the interview, phrase your questions carefully and deliberately. For example, by the way you phrase your questions you might be unconsciously setting the child up for certain types of responses. "Who moves the clouds?" "What moves the clouds?" and "How do you think the clouds move?" are questions based on three different assumptions. None of these is necessarily a bad question; the only inherent problem would be your not knowing exactly what you are trying to discover about the child's thinking. What particular aspect of the child's concept of the world—of nature, human life, or whatever—are you looking for? Ask your questions accordingly, and then *listen* carefully, without prejudging and especially without overestimating or underestimating what the child says. If a child has previously been diagnosed as "exceptional," try not to be swayed by the connotations of that label. In order to achieve better clarification or understanding, sequence ensuing questions so that they relate directly to the child's previous responses.

What sorts of questions should be asked, and to whom? The great advantage of this procedure is that you may ask whatever you are interested in, and you may ask whomever you wish. Piaget directed his questions toward all kinds of areas: the names of things, how or why things move, moral issues, the origins of things, causes of events. If you are interested in studying a fascinating collection of Piaget's investigative questions and examples of children's comments, read *The Child's Conception of the World* (Piaget, 1965). Ronald G. Good also has some interesting interviews in his article "Conversations with Children: Their Interpretations of Causality" (1972, pp. 193–196). Good directed his questions mainly to children from lower socioeconomic backgrounds. I once asked some middle-class suburban children from kindergarten to grade 5, "How do trees grow?" The following are representative, but certainly not all-inclusive, responses:

Kindergartener: "Because they're supposed to."

First-grader: "Because they're in the park."

Second-grader: "Because God makes them grow. He keeps pulling them out of the ground."

Third-grader: "So that squirrels can play in them. I like to climb them, too."

Fourth-grader: "Because they have to keep getting bigger, like us."

Fifth-grader: "I think it's because of photosynthetics or something like that. . . . You know what I mean!"

Try to interview the greatest variety of children that you can. You could look for variances in factors like age, sex, socioeconomic level, size of family, geographic setting, race, and so on. You might vary the themes of your questions or ask basically the same questions. It would be especially interesting to observe reactions to questions dealing directly with science, scientists, science teachers, and the learning of science.

Five Types of Reactions   The more children you are able to interview, the more commonalities you will eventually discover both in types of responses given and in ways children react to the questions. After analyzing the results of thousands of interviews, Piaget has categorized children's reactions to interview questions into five different types: random answer, romancing reaction, suggested conviction, liberated conviction, and spontaneous conviction (1965, pp. 10–32).*

Random answer   The child shows no interest, effort, enjoyment, or thought regarding the question. The answer is the first thought that comes to mind and is often flippant, noncommittal, unimportant, and consequently unstable. For example, "Trees grow because I say so."

Romancing (or fantasizing) reaction   The child shows little thought or conviction, but there is some evidence of enjoyment. The answer is meant to amuse the child or the interviewer. The child doesn't really believe it but at least has fun telling it. Romancing answers tend to be systematized and unique and show no general trends. For example, "Trees grow because the leaves are pulling on the branches, and birds push the branches, and bugs sometimes get in the trees, and our tree got broken by ice once, and . . ."

Suggested conviction   This is a reaction of the type "anything to please the teacher." The child tries to give the answer he or she believes the interviewer wants to hear. The answer often has the tone of a question. Since little thought

*From Jean Piaget, *The Child's Conception of the World*, Littlefield, Adams and Company, Totowa, N.J. Used by permission of Humanities Press, Inc.

or personal conviction is involved, a countersuggestion or even a skeptical glance from the interviewer can cause the child to modify the original answer. For example, *Child:* "Trees grow because they eat the soil." *Interviewer:* "Then trees have a mouth?" *Child:* "No. Well, sort of . . . Maybe it's because they already have food in them?" (Pause while the child searches for clues from the interviewer.) Finally, *Child:* "Oh, I don't know!"

Liberated conviction   This is a reflective, original answer, freed by the child for the first time as a result of the stimulus—the interviewer's question. It is as if the thought has been locked in the child's mind, and the question serves to liberate it. For example, "I think trees grow by adding more stuff like branches and leaves. I guess it's something like with us; they take in stuff and get bigger."

Spontaneous conviction   The child does not need to reflect but merely recalls the answer from previous experiences. The thinking originally required by the question has already occurred sometime before the question was asked. For example, "Trees grow because they keep adding new parts. I read a book on trees."

As you conduct your interviews with children, try to be aware of the kinds of reactions they exhibit toward your questions. Do certain kinds of children tend to react similarly? How do the reactions of an insecure or handicapped child compare with those of a well-adjusted or healthy child? *Don't try to equate the rightness or wrongness of an answer with the type of reaction.* These categories deal only with how the child reacts to your questions. An answer might be completely incorrect according to adult standards but still be a liberated or spontaneous conviction. Some reactions may be difficult to categorize. You may need to ask additional questions, like, "Have you ever thought of this before?" or "Did you study this in school?" You'll probably become more accurate with more practice and wider experience. But gaining accuracy in categorizing the reactions is not the primary goal. The real purpose is to gain greater insight into how children think or how they view their world. As an educator, you should be able to recognize when children are revealing their most personal, honest, original thinking (regardless of "correctness"), and when they are simply trying to please you or are indifferent or unchallenged.

It would be advantageous to organize your interviews so that the results could be shared and compared. You might use the format suggested in Table 3-2, or you might develop your own system. It might also be helpful to refer to Table 3-1 as you interpret the results of your informal conversational interviews with children. These interviews can be interesting learning encounters for both you and the children. They will also serve as excellent preparations for the next technique, the use of conservation tasks.

Interviewing: Administering the Conservation Tasks

This third technique for identifying children's thinking pertains most directly to the characteristics discussed in Chapter 2. This is the clinical interview technique, in which Piagetian-type conservation tasks are administered. (The original tasks

Table 3-2
RESULTS OF INFORMAL CONVERSATIONS WITH CHILDREN

Child's name: _____ Age: _____

Background information (physical characteristics, sex, language patterns, geographic setting, socioeconomic background, etc.):

Question(s):

Response(s):

Reaction type(s):

designed by Piaget tend to be much less structured than the examples you will soon be considering. However, it should be easier for you to learn the tasks, administer them to children, and interpret the results if they are more structured initially.)

The purpose of this third model is more specific than that of free-flowing observations of or informal questioning of children. The Piagetian-type conservation tasks provide one of the best means of determining a child's intellectual stage of development. Some nightly television programs end with the question, "Do you know where your children are?" You have the same question to face with each new group of students—not in terms of their physical location (although in some instances, that, too, can be a problem), but in terms of their intellectual development. You simply cannot assume that all 9-year-olds are in the concrete-operational stage, or that all second-graders are preoperational thinkers, or that sixth-, seventh-, and eighth-grade students are formal thinkers. But after administering and interpreting a child's performance of Piagetian-type conservation tasks, you will be in a much safer position to make assumptions. One of your most important functions as a teacher is to diagnose and make decisions on the basis of your diagnoses. The information gained from interviews on Piagetian-type conservation tasks can serve as part of the basis for your decisions. The better you know a child's level of functioning, the more wisely you can select appropriate encounters.

Examples of Piagetian-type Tasks There is no collection of tasks "certified" as *the* official device for measuring children's positions in Piaget's intellectual stages of development. The activities which follow are presented as samples which can be used as presented or modified to meet special needs or interests. Two examples for each conservation concept will be given, so that (1) you'll be able to be more selective; (2) you can draw from more ideas to eventually develop your own tasks; (3) you can, if you like, administer one set of tasks at the beginning of the school year and the other later on; (4) if you are unsure of a child's initial performance in one situation, you can immediately administer a parallel task to gain more information.

Child

You

Figure 3-1

Figure 3-2

Figure 3-3

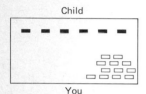

Figure 3-4

1. Conservation of number, example I: Checkers

Materials needed: Six black checkers and at least eight to twelve red checkers (the red ones are shown as white in our illustrations).

Procedures A: Distribute the black checkers evenly in a straight line. Keep the red ones clustered at one side (see Figure 3-1).

Request A: Ask the child to line up the same number of red checkers as there are black ones. (Then remove the unused red checkers so that they will not distract the child.) See Figure 3-2.

Special caution: If the child cannot demonstrate one-to-one correspondence, *do not continue this task.*

Procedures B: Do not move the black checkers. Stack the red checkers into a column as in Figure 3-3.

Request B: Ask the child the following questions: "Are there less red checkers than black checkers? Are there the same number of red checkers? Or are there more red checkers than black checkers?" *Ask the child to explain his or her answer.*

Procedures C: Move the black checkers into a tight line. Spread the red checkers as far apart as you can (see Figure 3-4). You might even spread the red checkers across the length of the room.

Request C: Ask the child: "Are there now more red checkers than black checkers? Are there less red checkers than black checkers? Are there the same number of red checkers as black checkers?" *Ask the child to explain his or her answer.*

Some suggestions: For a third configuration, the red checkers could be clustered close together in no particular pattern. Repeat the same questions as in request B or C. Notice the slight variations in the wording of these two sets of questions. It would be wise to vary the sequence often; in other words, don't always ask questions in the same order ("More? Less? The same?"). Also, remember *always* to ask for a justification after the child has made a choice.

2. Conservation of number, example II: Arithmetical relationships

Materials needed: Sixteen raisins or grapes or the like—anything that could be eaten by the child after completing this task.

Procedures A: Explain that you are giving the child four raisins for a morning snack and four for an afternoon snack. There are enough raisins for you to have two snacks, too. Arrange the raisins as in Figure 3-5.

Child

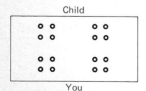

You

Figure 3-5

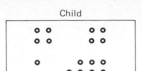
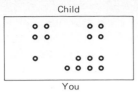

Child

You

Figure 3-6

Special caution: Ask the child, "Do we both have the same number of raisins (or whatever) to eat?" Again, if the child cannot affirm one-to-one correspondence, *do not continue this task.*

Procedures B: Explain that you want to eat only one raisin in the morning and save the rest for the afternoon. Move your raisins accordingly as in Figure 3-6.

Request: Ask the child: "Will you be eating more raisins than I? Will we both be eating the same number of raisins? Will you be eating less raisins than I?" *Ask the child to explain his or her choice.*

Some suggestions: This task deals with the relationship between 4+4 and 1+7, which requires a higher level of thinking and understanding than the first example. We will be dealing with the implications of these tasks for education in a later section of this chapter. However, try to formulate your own views on possible implications as you proceed with the conservation tasks. More specifically, if you happen to observe a first- or second-grade child "fail" this particular task, try to envision how that child must feel when confronted with an entire page of problems like this example.

3. Conservation of substance, example I: Solid amounts with clay

Materials needed: Two equal balls of clay, each about the size of a golf ball. (The two balls of clay could be of different colors for easier identification.)

Procedures A: Let the child handle the clay balls and, perhaps, heft them to confirm that they have the same amount of clay. If the child insists that they contain different amounts, encourage him or her to do whatever is necessary to make them the same.

Special caution: The child must believe that the two balls contain the same amounts of clay before one is altered. However, be careful not to overemphasize the sameness, lest you prompt or suggest later responses.

Figure 3-7

Procedures B: Ask the child to roll one of the balls into a long snake as in Figure 3-7. (If the clay is rather hard, you might have to assist the child.) Do not alter the other ball of clay.

Request: Ask the child: "Does the snake now have less clay than the ball? More clay? Or do the snake and the ball have the same amounts of clay?" *Ask for a justification.*

Some suggestions: When you ask the child to justify his or her choice, be careful that you don't overwhelm or frighten the child. To be confronted with a curt, "Why?" can be alarming to anyone, and especially to children. You might try a variety of approaches: "Why do you think . . . ?" or "That's interesting, but I wonder if you can tell me why you think . . . ?" Since it is the child's reasoning about the justification that is most important, handle this component of every task with great care.

4. Conservation of substance, example II: Liquid amounts with colored water

Materials needed: Equal amounts of colored water in two identical short, wide containers (like baby-food jars) and one tall, slender container (like an olive bottle or a graduated cylinder). Also, have some paper towels in case of spillage.

Figure 3-8

Procedures A: Have the child confirm that there are equal amounts of water in the two short, wide containers. (See Figure 3-8.) Again, if the child disagrees, encourage the necessary changes for making the amounts equal. (Equality must be initially established.)

Figure 3-9

Procedures B: Ask the child to carefully pour water from one of the short containers into the empty container. (See Figure 3-9.) If necessary, hold the base of the tall container for the child, to prevent spillage.

Request: While holding the tall jar, ask the child if there is more water than in the now-empty short jar, or less, or the same amount. *Ask for the child's justification.*

Some suggestions: Test the holding capacity of the tall jar before administering this task. Have enough colored water in the short, wide container so that when it is poured, it fills the tall, slender container almost to the top. Colored water is recommended simply because it is easier to see.

5. Conservation of length, example I: Distance traveled

Special caution: Be aware of the possibility that not all young children will understand the specific terms *length* and *distance*. However, they still might understand the concepts.

Materials needed: Two strips of paper, identical except for color, each 12 in. long and 1 in. wide. Any two colors can be used—for example, red and white.

Procedures A: Hand the child the two strips and ask him or her to show you if they are the same length. (You might ask instead, "Is one strip as long as the other? How can you show me?") If the child cannot demonstrate equality of length—for example, by superimposition or equal alignment—do not administer this task.

Procedures B: Arrange the strips as in Figure 3-10 and ask the child to use his or her imagination or to "make believe." Explain that the red strip (shown as black in Figure 3-10) is a sidewalk showing how far the child walks to school. The other strip (white in Figure 3-10) shows how far you walk to school.

Request A: Ask the child if you both walk the same distance to school. (Terminate the task if the child is confused.)

Procedures C: Move your strip about 6 to 10 in. ahead of the child's strip, as shown in Figure 3-11.

Request B: Ask the child, "Do you have less distance to walk than I? Do we both have the same distance to walk? Or do you have more distance to walk than I?" *Ask for the child's justification of his or her answer.*

Some suggestions: You might also try extending your strip even farther ahead of the child's until the entire strip is 1 to 2 in. beyond the child's. Watch for consistencies or inconsistencies in the child's responses. This task can also be modified by substituting small cars which the child is to imagine as traveling at the same speed along the strips. Which has more length to travel? . . . less? Or do they have the same length to travel?

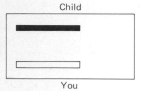

Figure 3-10

Figure 3-11

6. Conservation of length, example II: Lengths of patterns

Materials needed: Seven pieces of yarn of varying lengths; three of the seven should be of equal length. (For example, you could have separate lengths of 2, 4, 6, and 12 in. and three lengths of 10 in.)

Procedures A: Randomly place the pieces before the child, as in Figure 3-12. Ask the child to pick out the pieces of yarn which are the same length. Then remove the four unequal pieces. (Do not complete the task if the child cannot identify the three equal lengths.)

Procedures B: Place one of the three equal pieces in a straight line in front of you. Then make any kind of open, wiggly pattern from one of the other pieces. Ask the child to use the third piece to make another design. Figure 3-13 is one possible combination.

Request: Ask the child, "Which piece is the longest? Which is the shortest? Which pieces are the same length?" *Ask for justifications.*

Some suggestions: If readily available, different-colored yarns would facilitate reference to the different patterns. Pipe cleaners or thick but pliable wire could also be used instead of yarn.

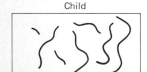

Figure 3-12

Figure 3-13

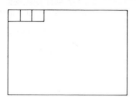

Figure 3-14

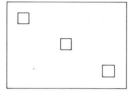

Figure 3-15

7. Conservation of area, example I: Grass-mowing problem

Caution: Again, a child may have the concept of area but not recognize the specific term *area.*

Materials needed: Two sheets of green paper of equal size (approximately 8 by 11 in.) and six identical cubes (about 1 by 1 by 1 in.) or six identical pieces of paper each 1 in. square.

Procedures A: Show the child the two pieces of green paper and ask if they are the same size. Also, ask the child to determine if the six squares are all the same size. (As you've probably guessed, equality must be verified first.)

Procedures B: Ask the child to pretend that each green paper is a field of grass. Place three of the cubes or squares on one of the green papers, as in Figure 3-14. Explain that these are houses built on the field.

Procedures C: Then arrange the other three cubes or squares on the second sheet, as in Figure 3-15. Explain that these are three houses built on the other field. You and the child have agreed to mow the grass in the two fields.

Request: Ask the child, "Will we have the same amount of grass to mow in both fields? Or will we have more grass in one field and less in another?" *Ask for the child's justification.*

Some suggestions: Modify this task according to the child's background. (In fact, that's a good policy to use for *all* the tasks.) For example, an urban child might be more familiar with sweeping or surfacing a parking lot in which cars are parked in positions similar to those of the cubes or squares in Figures 3-14 and 3-15. A rural child might relate better to grazing land for livestock, with the cubes or squares standing for barns.

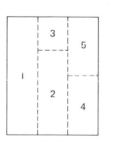

Figure 3-16

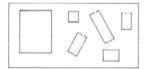

Figure 3-17

8. Conservation of area, example II: Paint job

Materials needed: Two plain sheets of white paper (regular 8½ by 11 in. will suffice) and scissors.

Procedures A: Have the child verify the equal size of the two sheets of paper.

Procedures B: Explain that you need to make something from the papers. Take one of the pieces and begin cutting it into five separate pieces, as shown in Figure 3-16. As you continue cutting the paper, explain that you really need one sheet to be painted red and the pieces from the other sheet to be painted blue.

Procedures C: Place the uncut sheet of paper near the child and randomly arrange the five cut pieces near it, as in Figure 3-17. Again, remind the child that the uncut piece needs to be painted red and the cut pieces need to be painted blue.

Request: Ask the child, "Will you need more red paint (point to the large piece), or will you need more blue paint (point to the five cut pieces), or will it take the same amount of paint to color this red as to color these blue?" *Ask for justification.*

Some suggestions: Any colors can be used, and the second sheet of paper can be cut in any shapes—but do keep the cut pieces fairly simple so that if a child decides to demonstrate reversibility, it will not require an undue amount of time.

9. Conservation of weight, example I: Clay boats

Note: Actually, the scientifically correct term is *mass,* not *weight.* The distinction between these two terms will be explained more thoroughly in Chapter 4. For the present, you can refer to *weight* unless challenged by a child who has learned the correct terminology. This task and the following one are rather similar. After you have read through them, you'll need to practice the procedures. Only then can you decide if you're a better clay-boat floater or aluminum-foil sinker.

Materials needed: Two identical balls of clay (each about the size of a golf ball) and a

Figure 3-18

container (like a bucket, a large bowl, or an aquarium) at least half full of water. (Use an oil-base clay, so that it will be reusable.) Also, have some paper towels available.

Procedures A: Again, have the child confirm the equality of the two clay balls—this time by asking about their equal weights. Encourage the child to pinch off or add on a smidgen if necessary. (Have an equal-arm balance available for the scrupulous skeptics.)

Procedures B: Ask the child to watch carefully as you place one of the clay balls on the top of the water and then release it. Ask the child to describe what happened. (Send your clay to the Navy Department or Ripley's "Believe It or Not" if it does not sink.)

Procedures C: Then shape the other clay ball into a rather flat boat and gently place it on the water surface, where it will float. (See Figure 3-18.)

Request: Ask the child, "Do the clay boat and the clay ball weigh the same? Or does one weigh more or less than the other?" *Ask for justification.*

Some suggestions: Practice molding clay boats before administering this task, if you want to avoid that strange sinking sensation. Flatten the clay to about 1/4-in. thickness and form sturdy sides about 1/2 in. high. Try to mold the clay boat quickly, so that you don't lose the child's interest. If the child seems quite dexterous, have him or her do the initial molding, and then you can add the final touches if necessary.

10. Conservation of weight, example II: Aluminum sinker

Materials needed: Six squares or rectangles cut from kitchen aluminum foil, preferably the heavy-duty type. Two of the shapes must be identical—approximately 3-in. squares will do. The other four shapes should be similar but easily distinguishable from the two identical shapes. Also, have a container of water and paper towels as in task 9.

Procedures A: Randomly place the aluminum foil shapes before the child as in Figure 3-19. Carefully note the next request.

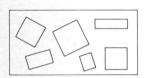

Figure 3-19

Request A: Ask the child to show you two pieces that *weigh the same;* do *not* rephrase the request by asking for same sizes. (In this situation, the child must assume that sameness of size equals sameness of weight. If the child cannot demonstrate this assumption, do not continue the task.) Put the unequal pieces aside.

Procedures B: Take one of the equal pieces and shape it into a boat by simply bending up the edges. (You could ask the child to do this while you are preparing the second piece.)

Procedures C: Take the other equal piece and carefully fold it in half, then in half again, and so on. Keep pressing the foil into as small and compact shape as you can manage.

Special caution: Try to avoid leaving any large air pockets in your folded foil. As a final touch, step on the piece so that it is as flat and compact as possible.

Procedures D: Place the aluminum boat on the water surface along with the folded, flattened piece. Ask the child to describe what happens. (See Figure 3-20.)

Request B: Ask the child, "Does one piece of aluminum foil weigh more or less than the other? Or do both pieces weigh the same?" *Ask for justification.*

Some suggestions: Develop your technique for sinking aluminum foil. As was mentioned at the beginning of these conservation-of-weight tasks, only after some practice can you decide if you're ready for clay-boat floating or aluminum-foil sinking.

Figure 3-20

11. Conservation of volume, example I:
Objects of equal volumes with different weights

Note: Since children do not usually acquire conservation of volume until their early teens, there seems little to be gained by administering either of these tasks to younger

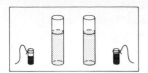

Figure 3-21

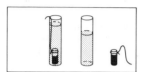

Figure 3-22

children or to the mentally retarded. However, if any child (even a precocious youngster) has demonstrated conservation of number, substance, length, area, and weight, it would be quite appropriate to administer these volume activities.

Materials needed: (This activity will require some scrounging.) (1) Two objects of equal volume (that is, the same size) but with easily perceptible differing weights. (Two identical narrow pill vials from a pharmacy could be used. Fill one of them half full with sand, pebbles, or any available weights and fill the other almost totally full.) (2) Two identical slender, tall containers (like olive jars) or graduated cylinders three-fourths filled with colored water. (The opening of these two containers must be large enough so that the equal-volume objects can be lowered into them.) (3) Equal lengths of string tied around each of the equal-volume objects. (4) Two small rubber bands fitted around each container at the level of the colored water. (See Figure 3-21.)

Procedures A: Ask the child to determine if (1) the identical containers have the same amounts of colored water, (2) the equal-volume objects are the same size, and (3) one of the objects is noticeably heavier or lighter than the other. Encourage the child to make any necessary modifications.

Request A: Ask the child to predict what will happen to the level of the water when the *lighter* object is lowered into the container. Have the child position the rubber band to show his or her prediction of the water level.

Procedures B: Use the string to gently lower the lighter object into the container. Compare the actual rise in water level with the predicted rise. Reposition the rubber band if necessary to correspond to the actual water level. (See Figure 3-22.)

Request B: Ask the child to now predict what will happen to the water level of the other container when the heavier object is lowered. Have him or her position the rubber band accordingly on the other container. *Ask for the justification now—* that is, *before* the heavier object has been lowered into the other container of water.

Procedures C: Concentrate on the child's reactions as the heavier object is lowered into the second container. Do not reposition the rubber band. Simply let the child observe and compare the actual rise in water level with the predicted level.

Request C: Ask the child to try to explain the results.

Some suggestions: Practice using the materials along with the suggested sequence of procedures and questions. Test that *both* the heavier and the lighter objects sink when they are lowered into the water, especially if you use the pill vials. (If the objects partially float, you will alter the intended concept to be demonstrated, namely, water displacement by volume.) Perhaps more than in any of the other task interviews, one is tempted to explain the results at the end of this interview. Please, avoid that temptation. The concept will be meaningful only if it is discovered by the child.

12. Conservation of volume, example II: Clay shapes with equal volume and equal weight

Materials needed: Use the same items as in task 11 but substitute two identical shapes of clay for the two objects of equal volume but differing weights. (The clay should be shaped so that it can easily fall into the container of water without being forced or sticking to the sides.)

Procedures A: Again, establish the equalities of (1) containers, (2) amounts of water, and (3) amounts of clay. (Have a balance scale available if necessary.)

Request A: Ask the child to predict what will happen to the water level when one of the

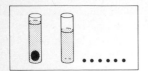

Figure 3-23

clay balls is lowered into the container. Have the child position the rubber band to indicate his or her prediction of the water level.

Procedures B: Lower one of the clay shapes into the container. (Be careful that the falling clay does not splash water from the container. Be ready to immediately cover the top of the container with your palm to prevent splashing.) Compare the actual rise in water level with the predicted rise and reposition the rubber band if necessary.

Procedures C: Form about six or more very small spheres from the other clay shape, as shown in Figure 3-23.

Request B: Ask the child to predict the water level if all the small spheres are dropped into the second container. Again, have him or her position the rubber band accordingly on the second container. *Ask for the justification now—before* placing the small spheres into the second container.

Procedures D: Complete the interview as in task 11, procedures C and request C.

Some suggestions: You can vary the shapes of clay in any number of ways. For instance, you could begin with sausage shapes and cut the second shape into many thin slices. Practice using the materials, especially so that you can avoid splashing water when the shapes are dropped into the containers. As in task 11, be sure that all the clay shapes are totally submerged when they are in the water.

The preceding twelve tasks have dealt with six types of conservation: number, substance, length, area, weight, and volume. You can gain additional insight into a child's intellectual development by administering tasks dealing with states of transformations. Here are two more examples of Piagetian-type tasks.

13. States of transformation, example I: Falling pencils

Materials needed: One lead pencil and a set of six index cards showing separate varying positions of a falling pencil. (See Figure 3-24.)

Procedures A: Hold the pencil in a vertical position as in Figure 3-24a and ask the child to observe it carefully. Then release the pencil so that it falls over freely to a horizontal position. (If the child seemed to be unaware of the actions, repeat the procedures one or two times.)

Procedure B: Place the six cards before the child in random order.

Request: Ask the child to arrange the cards so that they show how the pencil fell. *Ask for the child's justification.*

Some suggestions: In Figure 3-24, notice that the lower right corner has been clipped off each card. This can be a guide for keeping the orientation of the pencil tips constant when you randomly arrange the cards before the child. Otherwise the choices can become quite confusing. Encourage but do not force the child to arrange all six cards. The preoperational child will tend to identify the first and last events and to confuse or simply ignore many of the middle events.

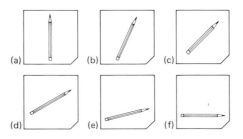

Figure 3-24

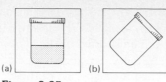

Figure 3-25

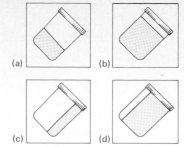

Figure 3-26

14. States of transformation, example II: Water level

Materials needed: A pencil and drawings of two bottles as in Figure 3-25. The bottle in Figure 3-25a has a tight lid and is about half full of water.

Procedures: Show the child a drawing like Figure 3-25a and ask the child to describe it. (If necessary, draw attention to the water and the tightened lid.)

Request: Then show a drawing like 3-25b and ask the child to draw how the water would look if the bottle were leaning or tilting. (If the child is concerned that the bottle will fall over, position your finger on the drawing as if to hold the jar in a tilted position.) *Ask for the justification.*

Some suggestions: It is sometimes difficult to phrase the question clearly for young children. They often envision only "spilt milk" when a container is leaning over. You may have to remind them several times that the lid is there to keep the water from running out. Preoperational children will tend to center on (1) the base of the bottle (Figure 3-26a), (2) the top of the bottle (Figure 3-26b), (3) the lower side (Figure 3-26c), or (4) the upper side (Figure 3-26d). The task deals with the child's conceptualization of the horizontality of the water level, not with the amount or volume of water. Therefore, accept any variations between the horizontal water lines shown in Figure 3-27.

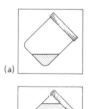

Figure 3 27

Preparations for Administering the Tasks The initial preparations are similar to those discussed in relation to the model for informal conversations. Establish a relaxed atmosphere and determine the most appropriate vocabulary level on the basis of your opening conversation with the child. Because of the nature of the interviews on the conservation tasks, however, some additional preparations need to be considered.

Materials Select your materials on the basis of suitability, safety, and attractiveness. Since every task requires interaction with things to be manipulated, package or prearrange your materials according to the order in which you wish to administer the tasks.

The child Interview only one child at a time. (Young children are especially influenced by others' answers and usually respond with, "That's what I was going to say.'') Have a purpose for selecting the child you wish to interview—for example, specific age level, socioeconomic background, level of assumed ability, physical attributes or impairments, or verbal level. Avoid interviewing your own children, relatives, or neighbors, since objectivity will be more difficult to maintain. If you select a child for a specific purpose, you will tend to give that child greater attention.

Observing   Various means can be used to record your observations: (1) Have someone else keep notes of the child's responses. (2) Tape-record the session for later playback and analysis. (3) If you must operate individually, jot down brief notes while you have the child either replace used materials or investigate new objects for the next task. (A suggested format for recording and analyzing observations is given in Table 3-3, page 60.)

Physical setting   Locate the best physical conditions for administering the tasks. Try to avoid a busy area where the child can be easily distracted. Select a comfortable setting which has ample workspace. If you will be interviewing more than one child from the same classroom, family, or neighborhood, try to limit their communication with one another.

Introduction   Decide how you want to initiate or introduce the tasks. Again, there are several options; select the technique with which you feel most comfortable and which best fits the child. You might explain that you have some interesting games to try out, but you are not quite sure if children will like them; ask the child to help you. Or explain that you have an assignment to complete and need the child's help. Children are generally amazed that adults have assignments to do and are most empathetic. Or simply explain that you have some materials and would like to ask the child some questions about them. (Never refer to the interview as a *test.* If necessary, convince the child that there are no trick questions. However, suspicion might mount if the child focuses on your note-taking.)

Pacing   Try not to rush. Give the child ample time to think and manipulate the materials. Terminate the interview if you notice fatigue, stress, or total indifference from the child. Plan on spending about 20 minutes to 40 minutes per interview.

Questions   Have your general questioning sequence planned in advance. Keep clarifying or rephrasing the questions until you are certain that the child understands your request. Try to match your vocabulary to the level of the child's.

Neutrality   Try not to praise; treat all responses neutrally. Instead of beaming over correct answers and frowning over incorrect ones, or saying, "Good" or "Hmm," you might simply repeat the child's answer and move on to the next procedure. Also, beware of the tendency to prompt the child or to use the interview as a "teaching" opportunity. (Understanding of the concepts involved

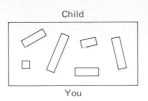

Child

You

Figure 3-28

in these tasks stems only from the child's interactions.) Sometimes the insecure, test-conscious child might persist with, "How am I doing?" Reassure the child but, again, maintain neutrality.

Check of basic concepts If you have any reason to doubt the child's understanding of the terms *more*, *less*, and *same*, the following procedures should be used *before* introducing any of the conservation tasks.

Materials needed: Six strips of paper, five of varying lengths, each about 1 in. wide. (For example: one 1 in. long, one 2 in. long, two 3 in. long, one 4 in. long, one 5 in. long.)

Procedures: Randomly place all six strips of paper before the child. Then pick up one of the 3-in. strips and tell the child that you will be using this one (see Figure 3-28).

Request: Ask the child, "Which piece of paper has more paper than mine? Which piece is the same as mine? Which piece has less paper than mine?" *Ask the child to explain all choices.*

Some suggestions: Try not to emphasize the words *more*, *less*, and or *same*. If the child wants to make direct comparisons with your 3-in. strip of paper, allow him or her to do so. After the initial choice has been made, you may also ask additional questions like, "Are there any other pieces that have more? . . . have less? . . . are the same?" Sometimes you might try challenging the child with questions like, "Are you sure?" "Really?" "What about this one . . . ?" You must be careful with this approach. It can be useful in helping you ascertain whether or not the child's reaction can be characterized as suggested conviction or liberated conviction (see pages 47–48). But it can also confuse some children and make them unduly nervous. You'll have to learn from experience when it is advisable or unadvisable to use countersuggestions. If the child is unable to distinguish which strips have more or less paper than your 3-in. strip, and which have the same amount, *do not administer any of the other tasks.* This concept check can be redesigned to use different materials in order to reassess understanding of the terms. Administering the other conservation tasks to a child who cannot correctly make these perceptual choices will be an exercise in futility.

How Are the Piagetian-Type Interviews Interpreted?

Organizing Data

By now you might be feeling as though you need the brain of a computer, the arms of an octopus, and the patience of a saint. How can one keep track of everything without getting lost or missing the forest for the trees? Examine Table 3-3 carefully. This table can serve as one means of organizing and categorizing your observations.

Keep the introductory information brief and pertinent. It can serve mainly to jog your memory after you have administered many tasks to many children. It can also supply you with data for making additional comparisons. The specification of materials used in the "Task" column should be simply descriptive and readily communicate which task you chose to administer. The letters M, L, and S under "Response" signify *more*, *less*, and *same*. The responses to the transformation tasks can be recorded simply as *yes* (Y) if the task was performed

Table 3-3
RESULTS OF PIAGETIAN-TYPE INTERVIEWS

Child's Name or Code_____ Age_____ Grade_____ School_____
Brief description of child:

TASK (SPECIFY MATERIAL USED)	RESPONSE	JUSTIFICATION (SUMMARIZE ANSWER TO "WHY?")	RESPONSE TYPES	REACTION TYPES (GIVE REASONS FOR CHOICES)
NUMBER	M L S		+ LN C R − CO CA NB	Ra Ro Su L Sp
SUBSTANCE	M L S		+ LN C R − CO CA NB	Ra Ro Su L Sp
LENGTH	M L S		+ LN C R − CO CA NB	Ra Ro Su L Sp
AREA	M L S		+ LN C R − CO CA NB	Ra Ro Su L Sp
WEIGHT	M L S		+ LN C R − CO CA NB	Ra Ro Su L Sp
VOLUME	M L S		+ LN C R − CO CA NB	Ra Ro Su L Sp
TRANSFORMATION (FALLING PENCIL)	Y N		+ LN C R − CO CA NB	Ra Ro Su L Sp
TRANSFORMATION (SKETCH STUDENT'S LINE)	Y N		+ LN C R − CO CA NB	Ra Ro Su L Sp

correctly or *no* (N) if incorrect choices were made. Circle the letter which corresponds to the child's ultimate response or final decision.

Under "Justification," try to record the child's answers to your requests for explanations as succinctly as possible. Without an accurate record of justifications, you will have no clear basis for selecting the appropriate items under "Response types." There are three ways in which a child can demonstrate positive (+) or correct logical operations in the encounters of the conservation tasks. The letters LN signify reasoning according to *logical necessity*. This is evidenced when a child responds or at least clearly implies that nothing was added, nothing was taken away; therefore, the result must still be the same. The realization "It *has* to be the same" should be evident. The letter C stands for reasoning by *compensation*. In our everyday use of the word *compensation*, we usually mean "making equal." For example, if you do some work for me, I'll pay you or compensate you. In that way we are both equal again. A child demonstrates reciprocal reasoning or compensation when one set of variables is equated with another. For example, if a child recognizes that the ball of clay is short and fat but the clay snake is long and thin, the two sets of variables offset and equal each other. The letter R is used to indicate that the child has demonstrated *reversibility* and can return the materials back to their original state. Must reversibility be physically evidenced, or can it be merely implied? I tend to prefer the former. When a child implies reversibility, gently persist with requests for clarification like, "I'm not sure if I understand what you're saying. Can you show me what you mean?"

When you conduct your interviews, strive to gather as much evidence as you can concerning the child's reasoning processes. For example, if a child immediately responds with some logical-necessity type of reasoning, reply, "That's an interesting answer. But I wonder if you can tell me anything else about why you think they are the same (or different)." Unless you are an extraordinarily patient person, you will probably have a tendency to rush your initial interviews or to center on only the classic answers or behaviors. But think about your own behavior when you are having a very personal conversation with a friend: you realize the importance of patience; you somehow know it's inappropriate to rush or force the conversation. Somewhat analogously, these interviews can provide meaningful moments with a child's thoughts. Thought processes seldom spring forth like a jack-in-the-box; neither should they be forced out. Allow them to unfold and evolve freely according to the child's initiative and inclinations. Then accept them nonjudgmentally.

Remember that the quality of a child's thinking is quite different from that of an adult. During your interviews you will most likely see at least three different ways in which a child can make responses that are illogical (by adult standards) or negative (−). The preoperational perception-prone child will tend to center on only one attribute, characteristic, or action at a time. The negative response type CO should be circled when the child "reasons" by *centering* on the *object* itself. For instance, the child might respond that the clay snake is more (or less or the same) because it is long or gray or thin or smooth. The child might also *center* on the *actions* (CA) which were observed. Some typical CA justifications might be, "It's more (or less, or the same) because I rolled it," or ". . . you poured it," or

". . . it was folded," or ". . . these were moved." The letters NB signify no basis for the choice, or simply a guess. This response also results when the child decides to repeat a choice that previously seemed to be successful. The child usually tends to hesitate more than usual, will search for clues from you, and may state the justification more in the tone of a question than as a declarative statement. There is usually a high correlation between an NB response to a task and a random answer or a suggested conviction.

Perhaps you've already decoded the items under "Reaction types." (If you need to review the five types of reactions previously discussed, see pages 47 to 48.) Briefly, Ra stands for "random answer," Ro for "romancing" or fantasizing, Su for "suggested conviction," L for "liberated conviction," and Sp for "spontaneous conviction." On your data sheet, simply circle the most appropriate choice, on the basis of reasons you briefly specify. It is highly possible that a child might exhibit several reaction types during the same task interview.

Interpreting Results

After sufficient data have been collected and categorized, you are finally ready to interpret the results. If necessary, review the characteristics of children at the different intellectual levels of development (see Table 3-1). Then study the evaluation form suggested in Table 3-4.

Information for the first five items of Table 3-4 can be derived directly from your observations as recorded in Table 3-3. However, I must caution you that the criteria developed for item 5 are based on my own personal experiences with children. Piaget did not develop the conservation tasks or related tasks as testing techniques for teachers or as diagnostic tools. Consequently, there are no concise guidelines available for interpreting the results.

My intention has been to provide you with some tangible Piagetian techniques for getting to know children better—how they view their world, how they interact with ideas and materials, and what kinds of thought processes they use. You were also cautioned that you cannot understand well either Piaget's ideas or children in general without firsthand experience or direct contact with children's learning encounters. Now you need to put some of Piaget's most basic ideas into practice. Administer at least six or more of the Piagetian-type task interviews to as many children between the ages of 5 and 14 as your time, interests, and energies allow. Then respond to item 6 of Table 3-4 as thoroughly and honestly as you can. Gradually you will be able to examine and evaluate Piaget's ideas more thoroughly and begin to determine which are most worth applying. The results of many interviews will probably stimulate the next question.

Why Are There So Many Differences among Children?

Even when you knew nothing of Piaget's ideas, you had probably noticed a great many differences among children. The greater the variety of your

Table 3-4
INTERPRETATION OF PIAGETIAN-TYPE TASK INTERVIEWS

CHILD _____ AGE _____

1. This child demonstrated conservation of _____ number, _____ substance, _____ length, _____ area, _____ weight, _____ volume. (Check the tasks which were performed logically.)
2. The states of transformation were recognized in _____ the falling-pencil task; _____ the tilted-bottle task.
3. The child exhibited the following:

PREOPERATIONAL TRAITS	CONCRETE-OPERATIONAL TRAITS

4. The rank order (first to last) of reaction types was:
   _____random_____romancing_____suggested_____liberated_____spontaneous.
5. The child's intellectual stage of development seems to be
   _____Preoperational (no conservation)
   _____Low transitional into concrete-operational (conserved one or two concepts)
   _____Beginning concrete-operational (conserved three to four concepts)
   _____Concrete-operational (conserved all but volume; succeeded with transformation)
   _____Transitional toward formal-operational (justified all tasks correctly).
6. As a result of administering and interpreting the interview, I learned the following about myself and about Piaget's ideas:

experiences with children, the fuller your realization of their differences. Administering the Piagetian-type task interviews can confirm and amplify this realization. But how do we account for these differences? More specifically, what factors influence a child's development? Is heredity the main factor, or environment, or a combination of the two, or what?

Piaget has never actively participated in the seemingly endless battle between the camps of heredity and environment. However, these factors are incorporated in his explanations. He has directed his answers to four rather obvious factors: (1) organic growth and maturation, (2) acquired experience, (3) social interaction and transmission, and (4) equilibration. But the real key to the answer seems to be something that is often overlooked, namely: (5) affective factors.

Organic Growth and Maturation

The importance of this factor is obvious to anyone who has had the jarring experience of visiting the section of a mental hospital where severely brain-damaged persons are kept. And even in less dramatic experiences, there is ample evidence that slow or abnormal physical development retards all other aspects of development—mental, motor, social, and emotional. There is a great deal that is not known about physical development. However, we do know that brain growth, the development of the central nervous system and the glandular systems, and improved motor, auditory, and vocal coordination (among other things) all open up vast new realms of possibilities for the child. Without the maturation of the physical abilities and the related structural capacities, the child simply would not have the wherewithal to interact with objects, events, or ideas. Organic growth and maturation are obviously important, then; but since they are normal in most people, the other factors have even greater influence.

Acquired Experiences

Acquired experiences are not the types of experiences reserved for the culturally enriched, such as the museum, the zoo, the planetarium, the symphony, and other wonders of the world. Rather, they are two basic types of experiences which any child has the opportunity to acquire: physical experiences and mental experiences. The latter are what Piaget calls *logical-mathematical* experiences. The former are prerequisite for the latter and have to do with perceptual growth. The child (and the adult) must have opportunities to physically interact with new objects or events in order to discover their physical properties, composition, and behavior—namely, their essential makeup. For instance, a child will never develop object permanency unless objects are often perceived as appearing, disappearing, and reappearing; and your understanding of the lunar surface can in no way equal that of the astronauts who actually bounced around on it. Physical experiences which promote perceptual growth are important. But in and of itself, perceptual growth is insufficient.

The child also needs the opportunity to *act on* objects or events, first physically and then mentally. It is the child's acting on real objects (for example, by arranging and rearranging objects, or by ordering and reordering, or counting and recounting) that evolves into logical-mathematical experiences. *This type of acquired experience deals with learning from acting on objects rather than learning about the objects themselves.* It is the logical-mathematical experiences which eventually become interiorized actions or operations that enable a child to conserve, for instance, and to override the tendency to judge solely perceptually. These experiences must be acquired by the *child* or the learner, not merely acknowledged or required by the parent or teacher.

Social Interaction and Transmission

The quality and quantity of verbal and social exchange also affect a child's development by providing opportunities to (in Piaget's phrase) *co-operate*—that is, to perform mental actions on objects or ideas with someone else. However, these experiences will be beneficial only to the extent that the child has the structure or capability to assimilate what is given and received. (For example, if you spoke no German, you would not immediately profit as much from living with a German family in Munich for a few weeks as would someone who spoke fluent German.) It was noted in Chapter 2 that language development is important to intellectual development but not totally instrumental. Logical thinking stems from the child's actions, not from the language received. Social interaction is probably the most powerful influence in weakening the tendency toward egocentricity, one of the strongest barriers to cognitive and social growth.

Equilibration

This is the active, self-regulating process whereby a child responds to encounters with external disturbances by coping or adapting in order to reach some tentative equilibrium. It's the interaction model diagramed on page 26. This is the mechanism whereby a child moves from states of lesser stability to greater stability, from incomplete understanding to more complete comprehension. It includes the component of disequilibrium or being "bothered," an element which is superseded by the child's need to change and to undergo an internal reorganization so as to achieve better equilibrium or the "I've got it" state. There is little doubt that equilibration is the most important and the most integrative of the factors discussed so far. If one has never experienced the self-regulatory process of equilibration, one has never experienced the joy of learning. I often feel that people who choose education as their profession probably have a special affinity toward, or realization of, the self-regulating nature of the equilibrium-disequilibrium-equilibrium interactive processes. I think they value the struggle and the eventual self-satisfying outcome of learning more than people not directly involved in education. They, I hope, can better recognize the self-motivating potential inherent in real learning. These beliefs are related to the affective factors, which have been ignored for perhaps too long.

Affective Factors

Piaget has been criticized for not including more data or insight on the roles that emotions—the affective factors—play in a child's development. It is easy to agree with these critics, but Piaget's position can also be defended. He certainly did recognize the importance of affective factors but chose, instead, to emphasize the empirical approach to his studies. Maier's study "The Cognitive Theory of Jean Piaget" (1965, pp. 88-89) stresses the inseparability and interdependence of intellect and affect: one directly affects the other; both are involved in the total adaptive process. Piaget and one of his colleagues, Inhelder, described the importance of the affective factors most forcefully in their book *The Psychology of the Child* (1969, pp. 157–158). They say, "In the final analysis, the affective factors are the key to all mental development." They go on to say that the real motivational forces and energizers for learning and intellectual development are affective drives such as the need to grow, to assert oneself, to be admired, to love, and to be loved. The presence or absence of these factors has the greatest influence on why there are so many differences among children and among all of us.*

Depending on the age or the "mental sets" of the children you will encounter in the classroom, your influence on the first four factors discussed may be minimal. The children's organic growth and maturation may be well established either normally or abnormally. They may already have been enriched by or deprived of the influences of acquired experiences, social interaction and transmission, and equilibration. But extremely few children are so hardened that they cannot still benefit from the affective factors. As you interact with children and provide encounters for them, how will the development of their self-concepts be enhanced or hindered? How will each child's affective drives be fostered?

No educators or psychologists know all the underlying processes involved in development and learning. But we can profit enormously from seriously considering at least these five factors which influence intellectual growth. The fact that Piaget's technique can help us better recognize children as unique individuals is just one of the implications to be derived from his theory. Some others should be examined, along with the ramifications of this one.

What Are Some of the Implications of Piaget's Ideas?

At the beginning of this chapter, you were urged to personally discover evidence to either support or refute Piaget's ideas by observing and analyzing children. I hope you have been gathering ideas and formulating opinions on the implications of Piaget's ideas throughout this chapter and Chapter 2. Your own discoveries will probably be far more significant than the summation offered here. The following comments are suggested not as a definitive compilation but as a basis for comparison and future consideration on your part. A threefold categorization will be used for what I consider to be the most pertinent implications of Piaget's ideas: (1) learners, (2) learning encounters, and (3) you, the educators. However, since all three of these components are involved in the integrative processes of learning and development, don't be too surprised if you find some overlap.

Learners

Piaget's theory strongly discounts the idea that children are simply miniature adults. The notion that children are like us in practically every way, except that they happen to lack knowledge and experience, can't hold water; and the "fill 'em up" approach to teaching is therefore empty and fruitless. Children differ in vast and varying degrees from adults in their perceptual abilities, language

*Used by permission of Basic Books, Inc., from Jean Piaget and Barbel Inhelder, *The Psychology of the Child* (Helen Weaver, trans.).

comprehension, and thought patterns. Children are not born with logic. A pedagogical policy is inappropriate and ineffective if it is built on the goals "to instruct" (from the Latin for "to build") or "to teach" (from the Anglo-Saxon for "to show"). Such goals imply that children already possess some sort of innate logical faculties; therefore, learning will occur simply if they are shown additional information or if the capabilities they have already formed are nourished and built up.

The characteristics of children, especially at the sensory-motor and pre-operational stages of intellectual development, certainly provide additional dramatic evidence that logic is not innate in the child. Logic and learning evolve gradually and are created or acquired by the child as a result of self-discoveries rather than as a result of what we teach. Each characteristic of the child at each level of intellectual development has important implications for the process of teaching and learning. It seems unnecessary and tediously redundant to review each characteristic, but perhaps a few reminders might help at this point.

You'll be wasting your time and energies when you use adult language and logic to reason with a preoperational child or when you expect a concrete-operational child to learn from verbal or printed propositions rather than from personal interactions with objects or events.

You also need to realize that in a great many instances children learn more from each other than they do from us. Interaction and cooperation among peers (the acting on objects or ideas with other learners) are essential to learning. The policy expressed by "Be still and stay in your seat (alone)" is often detrimental to the learning process. Certainly, there must be rules and limitations so that others' freedom and actions are not impeded. But children can establish these themselves, with your guidance. Efforts to decrease egocentricity are well worth a certain amount of student-to-student verbal communication (or "noise") and cooperation.

Much to our astonishment, and sometimes to our embarrassment, children come to us (at any grade level) with an impressive repertoire of practical knowledge. What a mess curriculum "experts" would make of designing instructional units for "teaching" object permanency or temporal succession to a sensory-motor child, or various conservation concepts to a preoperational child, or space-time relationships to a fifth-grade ball player! You must constantly remember that real learning occurs in the learner. The learner must have opportunities for personal discoveries.

Learning Encounters

First and foremost, the learning encounter should be an invitation to activity. It should provide the learner with opportunities to participate in the two most fundamental firsthand experiences: the physical experiences of exploring the properties of objects and the logical-mathematical experiences whereby self-discoveries can be made. (For example, a 5-year-old could look at ten checkers all day, and the checkers *themselves* would never reveal the sum "10." The child must act upon the checkers in a variety of ways which eventually lead to counting before the sum of 10 is discovered.) Physical manipulations must precede internal or mental manipulations. The learning encounter should both facilitate and activate the child's self-regulating process; that is, it should encourage and allow the child to freely control his or her learning. *You*, for example, cannot cause a child to become a conserver no matter how diligently and ingeniously you work at it. Conservation concepts can be acquired only as a result of the child's own operative self-discoveries.

The learning encounter should not provide the child only with the opportunity to be active for the sake of activity or to release energy. It should also be of such a nature that it provides the child with a need or a reason to do something meaningfully. Therefore, the learning encounters you select should have at least some moderate quality of novelty yet be within the capability of the child. As was explained in Chapter 2, if the encounter is below or above the learner's present capabilities, the result will usually be either boredom (because it is too easy) or frustration (because it is too difficult).

As you consider possible learning encounters, you must constantly analyze and evaluate the appropriateness or inappropriateness of the concepts involved. You can do this best by again considering the characteristics of children in the various stages of intellectual development. Often, learning encounters are selected for such nonsubstantive reasons as, "That's what's in the textbook to be covered" and "That's what the curriculum guide says to do" and "That's the way it's always been done." It is often because of reliance on these sorts of reasons that many inappropriate concepts continue to be perpetuated in elementary science. Topics like the molecular makeup of matter, electricity as a flow of electrons, the causes of day and night, the position and motion of our planetary system, the geologic periods, the chemical formula for photosynthesis, and the replication of DNA are just a few examples of inappropriate learning encounters found in some elementary science programs. Because they are there "to be covered," they're often taught (that is, shown) to children. But that does not necessarily mean that they are learned by children.

Learning encounters for preoperational children or some mentally retarded children should include activities that involve observing, comparing, describing, measuring, recognizing symmetry or space-time relationships, designing simple classification schemes, or seriating real objects. These same types of activities can be provided for concrete-operational children, with the addition of activities that involve inferring, predicting, and collecting, organizing, and interpreting data. The learner who is approaching the formal-operational level can have learning encounters that involve controlling variables, formulating hypotheses, and defining concepts operationally. Real experimentation, finally, is most appropriate for formal thinkers.

One of the most important implications of Piaget's ideas in terms of the learning encounter is the recognition and implementation of the premise that thinking or learning results from *actions*, not from words. If you accept Piaget's characteristics of children at the different intellectual stages, then you can realize more fully that verbal or printed propositions or explanations are not meaningful operative activities until the learner is at the formal stage of development.

Therefore, one of the greatest needs of elementary science education is a *deemphasis* on reading and writing, especially in the primary grades.

If you honestly attempt to analyze the daily fare of children in the vast majority of elementary school classrooms, you'll have to admit that the success-achievement-reward system depends first on reading, second on writing, and third on arithmetic. The goal of developing children's thinking skills or their self-regulating learning abilities can barely be recognized as a distant fourth. Too often teachers have failed to realize that physical manipulations come before mental manipulations; thinking stems from actions rather than from words or symbols. Science as a noun (physical experiences) and a verb (operations, first physical and then internalized) is a natural learning vehicle for children. Therefore, one of the most potent implications is that science should no longer be the subject taught on Friday afternoon if the art projects are finished or if all the reading worksheets have been turned in. Rather, *science should be the primary means for facilitating thinking and learning*. Who knows—could it possibly be that schools themselves are the causes of reading disorders and learning disabilities because of their insistence on inappropriate content and skills at the wrong time in a child's life?

Educators

Perhaps some of the points suggested in Chapter 2 are more meaningful now—such as the need for teachers to change from knowledge dispensers to designers of learning encounters. Which encounters will be most appropriate? Which will arouse the most interest and stimulate the child's needs? You will not be able to answer these questions without becoming a more skillful diagnostician.

It is all too easy for us to forget what it was like to be a child. Adulthood can sometimes be disadvantageous; it can deaden our sensitivities, leading us to misinterpret children or to formulate false assumptions about them. Therefore, it might be necessary to make very conscious, deliberate efforts to become a better observer and listener in order to determine what a child can do and what a child is really saying. Otherwise our expectations can be totally inaccurate or at best relatively unrealistic. Analyze, for instance, how you or others react to the answers children give to questions, especially when an answer is incorrect by adult standards. Do you (or others) immediately let the child know, "That's wrong"? Or is an attempt made to understand how the child arrived at that particular answer? When the child is experiencing disequilibrium, are you (or others) sensitive about when to intervene and when to stay out? How many teachers really believe that when teaching takes the form of showing or telling, it deprives the child of the opportunity for self-discoveries and subsequent real learning? How many realize that a far better technique is to pose alternative questions or suggestions to help the child reassess his or her original ideas?

All this is simply a variation on the same theme: the need to recognize and allow for individual differences. Strive to gear your attention and your selected learning encounters more to individuals than to the total group or to that thirty-headed pupil called "the class." This implies giving children opportunities to make meaningful learning choices—not the teacher's "Do you want to complete this reading worksheet or this grammar usage exercise?" or the parent's "Do you want to do your homework or clean up your desk?" It's strange that children are often criticized for making poor choices, when they've actually had few opportunities to exercise or even develop the skills involved in decision making. Could it be that many educators are subconsciously suspicious or openly disbelieving of the child's self-regulating process?

One of the greatest benefits we could derive from Piaget's ideas is an increased faith in the child's capacity to learn. Perhaps the most obvious sign of this faith would be our acceptance of the admonition, "Slow down." It seems as if our society, including education, is increasingly geared for speed. Parents especially are often guilty of trying to force their children's development too quickly and of expecting accomplishments prematurely. All of us could better realize that assimilation and accommodation take time, that learning is an ongoing process, and that *total* understanding of just about anything is beyond the capability of a child (and probably of an adult, too). Were we to slow down, we might also recognize the fact that concept development occurs much more slowly than motor or language development. There are a great many events that cannot be forced or accelerated without inflicting damage—sometimes irreparably. What do you think might be the most dramatic evidence of a commitment to the belief in the child's capacity to learn as portrayed by Piaget? It seems to me that a rupture or at least a major reorientation would occur in the entire educational endeavor. The *child's learning* would receive far more emphasis than the teacher's (or parent's) teaching.

What Are Some Impediments to Applying Piaget's Ideas?

One phenomenon that we can always count on is change. Everything is subject to some degree of change. Almost equally reliable is the phenomenon of resistance to change. Resistance to change in general is probably the basic reason why Piaget's ideas have not been more widely accepted. But there also seem to be four other recognizable factors: parents, administrators, existing curricula, and other theories.

Parents

School boards and parents, usually rightfully, have a very strong influence on the education of children. But unfortunately parents sometimes take the attitude, "What was good enough for me (in school) should be good enough for my kids." Note the parenthetical addition of "in school," an addition which is somewhat of a paradox. Parents generally want many more material gains or advantages for their children than they had, yet when it comes to education, their generosity and liberality dwindle. Parents who benefited from their early education might well value their experiences; and because they have no basis for comparison, they feel that their children should have the same sort of education as they had. Parents who did not benefit from, or who were scarred by, their early schooling are not vocal and probably only silently hope that their kids get a better deal. Parents, too, are caught up in society's desire for speed, and they tend to become impatient when they see "no progress" in their children. To suggest that formal reading instruction be delayed until the child was in the third or fourth grade, or until the child was physiologically ready, or until the child indicated the desire to learn to read would shock many parents or fill them with consternation or panic. Unfortunately, some teachers, too, share these attitudes.

Administrators

Even before the advent of accountability or management by objectives, administrators were test-conscious and diligent about covering the text. Woe betide the teacher whose class showed up poorly on the standardized test scores; woe betide the administrator who did not yield to pressure from parents; and woe betide the class which the administrator had to quiet because the children were noisy (having social interaction), making a mess (acting on objects), or moving about too freely (making learning choices), or which couldn't name all the planets and give their distances from the earth (inappropriate concepts).

This picture might be exaggerated. There certainly are many excellent, highly supportive elementary school administrators. But numerous examples of the picture painted above also exist in our schools. The point is that administrators can hinder (as well as support) the application of Piaget's ideas. If you happen to have the luxury of choosing when applying for a teaching position or selecting a school for your child, try to determine how well the administrators support your educational beliefs. You might also try to learn how much red tape you would have to go through in order to effect changes in procedures and goals.

Curricula

Too often curriculum designers and publishers are oblivious of Piaget's contributions or just do not accept them. Inappropriate materials and concepts continue to be perpetuated. Many textbook writers continue to equate learning with language acquisition and mastery rather than as the thinking derived from the child's acting on objects, events, or ideas.

When you have some spare time, read *Teaching as a Subversive Activity* (Postman and Weingartner, 1969)—at least Chapter I, "Crap Detecting," Chapter IV, "Pursuing Relevance," and Chapter V, "What's Worth Knowing?" Even without reading the book, it would be helpful just to think about those chapter titles in terms of science curricula.

Disequilibrium: Other Theories

It would be unfair to you and narrow-minded on my part to end this chapter without recognizing the contributions of other learning theorists, educational psychologists, or curriculum specialists. Piaget's works are extremely thorough and well-suited to elementary science education. But they are not all-conclusive and definitive. Perhaps your own sense of disequilibrium and curiosity have caused you to wonder, "But is there more?" If this is so, you might also investigate the works of several other scholars.

David Ausubel (1963) has shed much light on meaningful verbal learning. He also explains the role and importance of ideational organizers—that is, the means of determining how to bridge the gap between what the learner already knows and what is to be learned next.

Jerome Bruner (1966) postulated in 1960 that any subject can be taught to anyone at any age, provided that it is taught in an intellectually honest and useful way. Bruner has also shown the importance of creativity, discovery, and inquiry in the processes of learning.

Robert Gagné (1970) has designed a cumulative hierarchy of learning which begins with the simplest stimulus-response learning and terminates with problem-solving ability. Gagné has also defined four levels of scientific inquiry which are quite applicable here: the competent performer, the student of knowledge, the scientific enquirer, and the independent investigator.

Carl Rogers (1969) strongly supports the belief in the child's capacity to learn and the role of the teacher as facilitator of learning rather than dispenser of facts. Rogers stresses personal development in terms of self-discovery.

Hilda Taba (1962) categorized intellectual processes into three cognitive tasks: concept formulation, interpretation of data, and application of principles and facts. Hers is one of the few theories that has actually been translated into curricular designs.

This list could certainly be extended but should suffice as a starter. You will have to decide whose ideas make the most sense to you and can best help you understand children and their learning. Only you can answer the question, "How can I apply Piaget's ideas—or the ideas from any other learning model?" I hope the ultimate effects of your decisions will be reflected in both your actions and those of your students; and that your ultimate decisions will enable you and your students to place greater value on the adaptive processes of learning, living, and loving.

Encounters

1. Share the results of your informal interviews with your colleagues. In what ways were the results similar? In what ways were they different?
2. How might the five types of reactions be used to improve one's teaching?
3. Share the results of your Piagetian-type conservation-task interviews with your colleagues. Again, look for similarities and differences.

4. Which ideas from Piaget did you consider most important *before* your interviews with children? Which did you consider most important *after* you interviewed children? How can you account for the changes in, or the consistency of, your ideas?

5. On the basis of your interviews with children while administering the Piagetian-type conversation tasks, to what extend did the children correspond or not correspond to Piaget's characteristics of children in the four stages of intellectual development?

6. Develop some additional tasks or activities dealing with the conservation concepts or states of transformation. Test your ideas with children and share the results with your colleagues.

7. Analyze several elementary science textbooks (along with the teacher's guides) for the appropriateness or inappropriateness of the concepts presented. In what ways do the concepts and activities support or repudiate Piaget's ideas on active learning by children?

8. Spend some time observing science lessons being presented to children. How many opportunities are there for self-directed learning as opposed to teacher-directed showing?

9. What do you consider to be (a) the most important implications of Piaget's ideas; (b) the greatest impediments to applying Piaget's ideas; (c) the most necessary and important changes needed in order to model science education more along the lines of Piaget's ideas?

10. What other questions or areas of interest related to this chapter do you want to investigate?

References

Ausubel, David: *Psychology of Meaningful Verbal Learning*, Grune and Stratton, New York, 1963.

Bruner, Jerome: *On Knowing (Essays for the Left Hand)*, Atheneum, New York, 1966.

Gagné, Robert M.: *The Conditions of Learning*, Holt, New York, 1970.

Good, Ronald G.: "Conversations with Children: Their interpretations of Causality," in Ronald Good (ed.), *Science—Children: Readings in Elementary Science Education*, Brown, Dubuque, Iowa, 1972.

Meier, Henry W.: *Three Theories of Child Development*, Harper and Row, New York, 1965.

Piaget, Jean: *The Child's Conception of the World*, Littlefield, Adams, Totowa, N. J., 1965.

————, and Barbel Inhelder: *The Psychology of the Child*, Basic Books, New York, 1969.

Postman, Neil, and Charles Weingartner: *Teaching as a Subversive Activity*, Delacorte, New York, 1969.

Rogers, Carl R.: *Freedom to Learn*, Merrill, Columbus, Ohio, 1969.

Taba, Hilda: *Curriculum Development: Theory and Practice*, Harcourt, Brace and World, New York, 1962.

Suggested Readings

Bruner, Jerome, Jacqueline J. Goodnow, and George A. Austin: *A Study of Thinking*, Science Editions, New York, 1967. (Studies on a variety of strategies used by people to learn concepts.)

Furth, Hans G.: *Piaget for Teachers*, Prentice-Hall, Englewood Cliffs, N.J., 1970. (Numerous techniques for developing thinking in the second half of the book; extremely useful classroom ideas.)

————, and Harry Wachs: *Thinking Goes to School*, Oxford University Press, New York, 1974. (A description of attempts at the Tyler School in Charleston, West Virginia, to put Piaget's theory into practice; also includes 179 exciting and stimulating thinking games.)

Ginsburg, Herbert, and Sylvia Opper: *Piaget's Theory of Intellectual Development*, Prentice-Hall, Englewood Cliffs, N. J., 1969. (An excellent, thorough overview of the four stages of intellectual development, plus one of the finest sections on the implications of Piaget's findings for education.)

Good, Ronald G.: *How Children Learn Science*, Macmillan, New York, 1977. (An elementary science methods text with extensive material on Piaget's theory of cognitive development, along with more ideas for conservation tasks and activities appropriate for science encounters.)

Kagan, Jerome: *Understanding Children*, Harcourt Brace Jovanovich, New York, 1971. (Extremely worthwhile background information on children's behavior, their thought processes, and the importance of the interactive motivational forces operating in a child's life.)

Matthews, Charles C., Darrell G. Phillips, and Ronald G. Good: *Student-Structured Learning in Science: A Program for the Elementary School Teacher*, Brown, Dubuque, Iowa, 1971. (A general elementary science methods text with very good emphasis on Piagetian-type interviewing techniques.)

Phillips, John L., Jr.: *The Origins of Intellect: Piaget's Theory*, Freeman, San Francisco, Calif., 1969. (A nontechnical, general summary of Piaget's theory; good follow-up reading.)

Piaget, Jean: *Science of Education and the Psychology of the Child*, Orion Press, New York, 1970. (Background on education since 1935; can provide additional insight on Piaget's principles and their implication for education.)

PART TWO

PART
TWO

ALL learning—like living—involves taking in stimuli and acting upon or responding to what is received. In the past, much emphasis has been placed on the assimilation or taking in of scientific facts or phenomena. Too few efforts have been directed at analyzing what the learner does to what is received in order to reach some degree of understanding. What forms of accommodation (or what change of viewpoint resulting from actions performed) must take place for learning to occur? In what ways can children be guided to respond to science encounters so that their learning will be more meaningful? What kinds of science activities are appropriate for children?

The four chapters in this section can do more than answer these questions. They will help you discover that the thinking skills used in adapting to science encounters have surprisingly broad applicability. You can begin to discover that children can use the same kinds of skills that scientists use. You, too, can use these skills to make more sense out of your world.

SCIENCE FOR CHILDREN
DEVELOPING SCIENCE SKILLS

CHAPTER 4
GETTING STARTED: OBSERVING AND MEASURING

While investigating the natural world, children can be guided to perform many of the actions that scientists typically perform.

At first glance, this statement may seem preposterous. Certainly, children lack the storehouse of knowledge that scientists possess. They rarely embark on the search for truth with any semblance of a systematic approach. They cannot handle all the elaborate equipment that scientists use. And in fact no noteworthy scientific discoveries by children have ever been acknowledged. There have been child prodigies in music, drama, and a few other fields, but not in science. Obviously there is an enormous difference between a child's "playing scientist" and *being* a scientist. How, then, can that opening statement be supported?

Several ideas from Part One need to be reconsidered. *First*, science is both the accumulation of facts about the natural world and the search for these facts. Children do indeed engage in the search for truth. They question, wonder, observe, compare, make guesses, test ideas, and formulate some tentative conclusions. Until they've been squelched, they are insatiably curious. (The same is true of scientists.) *Second*, children learn by interacting with objects and events. (The basic tenet of scientists is empiricism; and the basic content of science is "hard data" derived from observations.) *Third*, although most children cannot independently perform hypothetical-deductive reasoning or controlled experimentation or deliberately speculate and accurately distinguish reality from possibility, they can be guided toward such skills. The guidance they receive will depend greatly on your efforts, on your decision-making ability, on your selecting the most appropriate science encounters which emphasize activity by the learners rather than passivity.

What sorts of actions should children be guided to perform? No one really knows the complete answer to this question, but two actions which seem to be the most fundamental will be examined in this chapter: observing and measuring. Nine additional actions or thinking operations will be considered in Chapters 5, 6, and 7.

Why Is Observing So Important?

The entire scientific enterprise is built on the skill of observing. All attempts to gain information about objects or events begin with observing. If scientists were not keen observers, they would never notice discrepancies, inconsistencies, or even the regular patterns of occurrences. Without data obtained from observations, they would seldom be bothered enough to ask questions and search for better answers. Observing is also one of the most natural actions for children. They have not lost their sense of wonder, because their five senses have not yet become anesthetized. They respond very freely to sights, feelings, sounds, smells, and tastes—often eliciting surprise or fear from adults. As children go about observing, their avoidance mechanism ("What will people think?") is virtually nonexistent. Their openness to all life and living, their receptivity and ability to respond to their surroundings are what we need to rediscover if we are ever to be able to do science with them. All teachers could benefit by considering their responsibility to continuously foster children's responsiveness as observers.

Learning begins with a response to a learning encounter. If the learner is unable to "reach out" (either physically or mentally) and take in the encounter, the interactive learning process will be stymied. Most often, it is the skill of observing that triggers all other aspects of the learning process. In the broadest sense—in life as a whole—observing is also vital to *caring*—caring about other people, things, events, places; caring about our total environment. If you are unobservant, you will fail to notice presences, absences, changes, and their effects. Without such observations, how can you care? Observing is not only the most basic of all science skills; it could well be the most fundamental act of a thinking person.

Some Observing Opportunities

Observing is a skill that can always be improved. No one can ever say, "I have mastered the skill of observing" or "I no longer need data from the five senses." Generally, the more opportunities you have for observing, the more you will improve your use of this skill. The following activities can strengthen both your skill at observing and your awareness of the environment. Five opportunities are provided, so that you can make choices. You will also have opportunities to observe cooperatively with others. There's no particular sequential order of difficulty in this set of activities. Most, if not all, can be easily adapted to classroom use with children. All except the final activity ("A Sense Celebration") could be conducted outdoors. Each activity has the same basic format:

A. An *introduction* which explains the general purpose of the activity and ends with a challenge or a reason or reasons for doing the activity
B. *Materials* which are readily available or which can be easily purchased or constructed
C. *Procedures* which consist of the basic road map to reach the goal along with key questions to keep you alert and on the right track
D. *Extensions* to suggest possible modifications, side trips, or new directions

Before you begin examining the observing activities, a few additional ideas need to be considered. Adults often do not use the skill of observing in complete isolation. We seem to have developed the tendency to *compare* as we observe—that is, to identify similarities and differences. However, this is not as common with children, especially preoperational children who center mainly on one characteristic at a time. In the following activities you will often be asked to *both* observe and compare. However, if you later wish to modify any of the activities for use with young children, it would be wise to stress observing an object's physical characteristics first. (Such observations might include size, shape, color, texture, hardness, heaviness, composition, position, location, etc.) Later, present activities which involve the children in determining similarities and differences between or among objects. Don't assume that observing and comparing occur spontaneously and simultaneously in children.

I've Got Mine!

A. Introduction Many times we observe only general features and fail to concentrate on distinct details. How well can you identify the *unique* qualities of an object when it is compared with similar objects? (This is a cooperative activity for six to ten persons.)

B. Materials A paper bag containing a collection of ten to fourteen similar objects, such as pine cones from the same tree, leaves from the same branch, almost-identical rocks from the same streambed, or whole unshelled peanuts from the same bag. Have at least four more objects than the number of participants in your group. These will serve as spares for later comparisons.

C. Procedures

1. Empty the collection of objects from the bag onto a table or workspace on the floor. Examine each object carefully.
2. After thorough observations have been made, place all objects in the bag aga n. Then pass the bag around so that each participant can take out one of the objects. Leave the spares in the bag.
3. Now, get to know your object well. What are its distinguishing characteristics? Note its colorations, patterns, shape, size, texture, odor, relationships among parts—any unique qualities you can find. Jot down notes or make sketches of several of the most important distinguishing characteristics.
4. After you have fully examined your object, replace it in the paper bag.
5. Again, empty the total collection from the bag and identify your object.
6. When were you able to say "I've got mine"? What enabled you to know?

D. Extensions Discuss the following questions:

Who has the most unusual object?
Which objects are most similar?
Which objects are most different?

Step 5 of the procedures could be modified dramatically by keeping your eyes closed and using the "pass it on" format. One person could act as leader and hand each participant one of the objects to be examined and then passed to the next person. Objects would continue being passed until all participants have received what they feel to be their original one. Then the visual check can be made.

Properties Matrix Scavenger Hunt

A. Introduction   This activity can be used to emphasize any single sense or any combination of two senses. For example, center on the use of the sense of touch. Make a list of all the descriptive words you can think of that deal with touch. How good are you at scavenging for objects with the characteristics you listed?

B. Materials   A matrix with touch descriptors such as Table 4-1. (If you use this activity with children, vary the size and complexity of your matrix according to the ability of the children. That is, use a larger chart with fewer descriptors for younger children.)

C. Procedures   First find representative objects which best exemplify the *separate* descriptors: an object that is basically smooth, one that is warty, one that is pitted, one that is soft, and one that is hard. Record the name of each object in its appropriate space. Then find twenty other objects each having *two* qualities to correspond to the remaining spaces in the matrix. For example, find something both warty and smooth, something both pitted and smooth, and so on. Again, record the name of each object in the appropriate space. How good a touch scavenger are you?

D. Extensions   This activity can be treated as a game by imposing a time limitation or giving points for each object found. However, the main purpose is

Table 4-1
TOUCH MATRIX

	SMOOTH	WARTY	PITTED	SOFT	HARD
SMOOTH	(Smooth)				
WARTY		(Warty)			
PITTED			(Pitted)		
SOFT				(Soft)	
HARD					(Hard)

Table 4-2
TOUCH-COLOR MATRIX

PRICKLY			
WRINKLY			
ETC.	GREEN	ORANGE	ETC.

Table 4-3
INTERACTIONS MATRIX

WATER			
TEMP.			
SOIL			
ETC.	TREES	FERNS	VINES	ETC.

to increase observing skills; be careful, therefore, that the element of competition does not become self-defeating.

Any number and variety of matrixes could be developed for each of the five senses or each combination of two senses. For example, you might try combining colors and textures as in Table 4-2, or odors and tastes, or shapes and sounds. A more sophisticated matrix could center on interactions between environmental factors and plants (or animals) as shown in Table 4-3.

The following "Sense Census" gives a related but different direction to the activity:

1. Make a list of four to eight descriptive words for each of the five senses.
2. Then look for objects that correspond to each word or several of the words from the separate lists for each sense.
3. Next, look for objects that correspond to words from the list of *two* senses, then for objects corresponding to words from the list for *three* senses, then *four*; and finally the full *five*.

The Odd Couple

A. Introduction   Are any objects truly identical? Although that question might be argued endlessly by philosophers, we'll pursue it in this activity. This activity will also facilitate the perceptual development of touch along with tne ability to detect similarities and differences. Can you discover the most mismatched pair?

B. Materials   Almost identical or seemingly identical *pairs* of objects like the following:

A used and a rather new tennis ball
A golf ball and a ping pong ball
Different kinds of indoor Christmas tree bulbs
Different kinds of outdoor Christmas tree bulbs
Large empty thread spools
Medium empty thread spools
Small empty thread spools
Tablespoons
Soup spoons
Teaspoons

You'll also need a "touch-feel box" into which one of each of the pairs can be placed. The box should have a hole just large enough so that you can reach in and take out any one of the objects. Have another box in which the other objects can be visibly displayed.

C. Procedures

1. Select any one of the visible objects; then reach into the "touch-feel box" to find its match. (Use only your sense of touch.)
2. Continue this procedure until all objects have been appropriately matched.
3. Now try to find the most "perfect pair." Which pair is most alike? What are the reasons for your choice?
4. Which pair is the greatest mismatch—that is, has the most differences? How many differences can you discover between your odd couple? (I've seen twenty-six differences enumerated. Can you establish a new record?)

D. Extensions   Any materials can be selected to substitute for those items listed. Use fewer objects with more obvious distinctions for younger children, and the reverse for older or more capable students.

You might also try using only one kind of object, for example, pairs of a variety of shells or acorns. (If you do use pairs of different kinds of acorns, glue the bottom cups to the kernels, since the two parts separate rather easily when the acorns are bumped around.)

A Bug's Eye View

A. Introduction   This activity is especially well suited for the outdoors, although with some imagination it could be conducted indoors. It is a modification of Steve Van Matre's "Micro-Trails" in *Acclimatizing* (1974, pp. 80-81). The first part of the activity should be conducted individually. But your efforts should later be shared with others.

Because our heads are so far from our feet, we very often miss what's going on at ground level. Even when we walk along a well-marked nature trail, there's much that we fail to observe. Part of the problem is that we are often bombarded by too many stimuli. In this activity we'll concentrate on the quality of a relatively small trail rather than the quantity of the more conventional nature hike.

Suppose that you've just been transformed into a bug-size creature, and you've been asked to develop the most interesting nature trail possible for your bug-size colleagues. What discoveries can you make from a bug's-eye view? What are the highlights of your scenic route?

B. Materials   Lengths of string (kite twine is fine) about 10 to 20 yards long; ten to twelve lengths of brightly colored yarn or ribbon, each about 3 to 4 in. long; a hand magnifying lens; and a small paper sack or plastic bag for carrying the materials.

C. Procedures
1. Select an area with good potential for laying out your mini-trail of string. Take your time, and by all means get down on all fours to fully capture the bug's-eye view of the area.
2. Establish a suitable starting point. (You might need to fasten the string with a stick or a rock in order to keep it from moving as you crawl around surveying the most scenic route.)
3. Every time you come to a special point of interest which you think should be noted, tie one of the pieces of yarn or ribbon to the trail of string to mark the spot. (Hint: there's no rule among mini-trail engineers that specifies only straight-line routes.)
4. Once you have run out of string and yarn markers, survey your trail again to determine if you've selected the most scenic route possible.
5. Then offer your services as tour guide to your colleagues. Share your discoveries.
6. Who constructed the most scenic route? Why do you think so?
7. What new discoveries did you make at the bug's-eye level?
8. What special mini-world was revealed along your mini-trail (or others' trails) as a result of using the hand magnifying lens?
9. Collect all your materials, lest you unduly complicate a real bug's view.

D. Extensions   Instead of being a talking tour guide, be a silent traveling companion and let your colleague identify and comment on your marked discoveries. Discuss the similarities and differences among trails. Again, concentrate on the use of your senses. Determine which was your favorite point of interest along any of the mini-trails.

Instead of simply flagging the points of interest along the mini-trail, you could construct accompanying learning stations. Self-explanatory directions could be taped onto each string, designating the tasks for the next visitor to perform. Such tasks might include comparing one particular plant with another; finding evidence that other animals had been on or near the trail before; and incorporating activities which relate mathematics, language arts, social studies, or art to the skill of observing.

You could also gear this activity for more specific purposes such as trails that exhibit interactions, changes, the use of one particular sense, or any variety of continuums dealing with color (brightest to dullest), frequency (most common to rarest), size (smallest to largest), or personal appeal (prettiest to ugliest).

A Sense Celebration

A. Introduction   Familiarity may or may not lead to contempt, but it can easily lead to indifference or unawareness. We can often overlook many common objects or events when we allow our senses to become dulled. Try observing something as if you were experiencing it for the first time or as if you had to describe it to someone who had never experienced it. Try to become more aware of your senses; and, as you do, celebrate them and what they bring you.

B. Materials   A birthday candle, matches, an index card, and a small piece of clay about the size of a marble.

C. Procedures
1.   Don't light the candle yet; first observe it carefully—as if you were experiencing it for the first time or as if you were never to observe it again. How many observations can you make? Try to list ten to fifteen descriptions.
2.   Place the small piece of clay in the middle of the index card. Stick the base of the candle into the clay so that the candle will not fall over. (If you have long hair, use a rubber band to hold your hair back. Ask your students to do the same if they work around any source of fire. You'll never regret taking this precaution.) Then light the candle, observe it carefully as it burns, and try to add ten to fifteen more observations to your list.
3.   You've observed the candle before and during its burning. Now, how many observations can you make after the burning has been completed?
4.   Take some time to examine your record of observations. Then code your observations in terms of your use of your five senses. Here's an example which you might use: label the observation SE if it is based on seeing, TO for touching, H for hearing, SM for smelling, and TA for tasting. You can use C if the observation includes a description of a change which you observed happening naturally or which you caused to happen.
5.   Which sense did you use most? Least? Which (if any) of your results surprised you?
6.   Compare your data with those of others. See if you can record thirty to fifty observations. (Then try to discover at least ten more.)

D. Extensions   Check the regulations of your school before trying this activity in a classroom setting. You might need to substitute other items for the birthday candle: for example, a "fizzy" tablet or small amount of food coloring dropped into water; a balloon before, during, and after inflation; or popcorn before, during, and after popping.

   Often it's very difficult to objectively consider the quality of observations. Being a good observer depends on more than just the quantity of observations recorded. One means of evaluating the skill of observing is to develop some sort of weighting system. The various values of the weights will depend on the factors you wish to emphasize or reward. Here's an example of one arbitrary system which you might try:

1 point for seeing.

2 points for touching.

3 points for listening.

4 points for tasting (if tasting is a safe procedure).

5 points for smelling (if safe).

2 points for recognizing a change which occurred naturally or which you caused.

3 points for quantified data (measurements).

Deduct 5 points when interpretations are made instead of observations or names of materials are used instead of physical properties (American Association for the Advancement of Science, 1968, p. 39).* (For example, the statement, "The candle is made of wax and a wick," is an interpretation lacking empirical evidence. "The cylinder consists mainly of wax-like material surrounding a narrow center of protruding twisted fibers" is a statement based on observations.)

Consider how you use your senses individually or collectively. Which sense or senses do you use most? Which do you use least? How would your life be different if you lost your sense of smell? Touch? Sound? Sight? Taste? Which sense do you value the most?

Additional Considerations on Observing

How would you have described the skill of observing before your encounters with the previous activities? If you are like most people, your description would have included only the use of sight. I hope that by now your ways of observing have been broadened. If you've become a more skillful observer, you will tend to (1) use all five senses whenever it is possible and safe, (2) note changes which either occur naturally or have been interjected, (3) quantify the physical characteristics which can be measured, and (4) refrain from giving interpretations or labels not derived from any of the five senses.

The previous activities have dealt mainly with the first two actions a skillful observer uses: observation itself, and comparison. We now need to consider more carefully the third element—measuring.

Why Is Measuring (in Metrics) So Important?

Observations are much more meaningful or at least more understandable if they are quantified. No one can know *exactly* what you mean when you describe a burning candle as "small," "short," "thin," "not very heavy," "emitting a fair amount of heat," and "disappearing moderately quickly." Measurement is needed to make communication clearer and more precise. Think about going to a grocery store lacking the skill of measuring, or ordering a "big piece" of

*Reprinted with permission from "Science" in *Science—A Process Approach, Commentary for Teachers,* 3d experimental ed., American Association for the Advancement of Science, 1968.

new carpeting, or building new bookshelves by "guesstimating," or trying to install a dishwasher in a space too small for it. How many arguments among children, especially the more egocentric ones, could be settled simply by their using the skill of measuring? ("She has more." "No, he has the most." "Mine is the fastest." "This is heavier than that." "Today is hotter than yesterday was.")

One of the major responsibilities of scientists is to make public disclosure of their findings so that their efforts can be verified, modified, or refuted. The more precisely they communicate their data, the easier the task of replicating their work. The measurements used universally in science are based on the metric system. In fact, the metric system is the official system of measurements in over 90 percent of the countries of the world. The United States is the *only* major country which presently is nonmetric. Perhaps you've seen bumper stickers saying "GO METRIC" or "THINK METRIC" or "METRICS ARE COMING." Where did the metric system come from, and why are we being urged to convert to it?

Some Background on the Metric System

The need for some kind of single, universal, comprehensive measurement system has probably existed ever since one person decided to measure with a favorite stick, another with a foot's length, another with a pace, and another with the distance from the nose to the tip of the middle finger. The situation must have been similar to that found in a typical preschool or kindergarten class—one child measuring with straws, another with erasers, another with paper-clip chains, another with broomstick handles. No one could agree with anyone else's results. Communication and common understanding were virtually nonexistent. During the 1600s and most of the 1700s, various proposals for standardization were recommended. But there was little agreement, and the measurement mess continued until 1790. It was then that the National Assembly of France formally asked the French Academy of Science to devise a standard for all known measurements. The ultimate result was the creation of the metric system, from the Greek *metron,* "measure." The physical standard chosen was called a *meter.* It was derived from a portion of the distance from the North Pole to the Equator. More specifically (even though the original calculations were inaccurate), the meter was equivalent to one ten-millionth of a meridian. For the first time, an invariable standard and a decimal system were established from which all other measurements could be derived. Acceptance of the new system was gradual but steady during the 1800s. By the 1900s, most of the world had accepted the metric system.

A major improvement in precision was officially adopted in 1960 at the Eleventh General Conference on Weights and Measures. Since the basis for the physical unit of the meter was discovered to be inaccurate, the officially accepted standard for the length of a meter became the wavelength in a vacuum of the orange-red line of the spectrum of the element krypton-86. (The exact specification of the meter is now 1,650,763.73 wavelengths.) This measurement is precise. It can be replicated infallibly under proper laboratory conditions.

During the 1960 conference, the metric system was also officially renamed *Le Systeme International d'Unites*, "The International System of Units," internationally abbreviated simply as SI. (The United States' present system has had to be unofficially renamed, too, from the "British System" to the "Customary System," since even Britain adopted the metric or SI system in 1965.)

Some Advantages of the Metric System

Any or all of the following reasons might be given in support of the metric system.

First, since the metric system (the *International* System of Units) is virtually universal, we need it to keep up with progress and to compete in world affairs and trade.

Second, if we were able to begin anew, we would quickly realize that the metric system is vastly easier than our customary system. The need to learn a litany of entirely different terms for related units would eventually disappear. (And who would really miss the old cup, pint, quart, gallon, peck, and bushel; dozen and gross; inch, foot, yard, and mile; ounce, pound, and ton—or their cumbersome fractions?) The metric system is far more logical; the units are derived from natural phenomena rather than from illogical and unrelated bodily proportions.

Third, since the metric system is based on ten, as is our monetary system, it is far less cumbersome and much more accurate to use. When converting from unit to related unit, one simply moves a decimal point to the right or left. The mental contortions required for conversions in our present system are entirely unnecessary.

Fourth, the fears associated with converting units from the customary system to the metric system or vice versa are completely unfounded. The two systems are like two totally different languages. To communicate, even if you are bilingual, you use one language or the other, as appropriate to the audience. Also, most of the necessary conversions will already be worked out for you—examine your grocery items more closely, or look at the wide variety of handy conversion tables that are available. Even gentle regrets ("But I like to teach inches, feet, pints, quarts . . .") need not be totally silenced. For quite some time, *both* systems will coexist and will need to be learned. Measurements in science-related lessons should be only metric, but curriculum writers in mathematics will simply need to do a reversal—instead of designating a few pages for metrics, they can include the customary units as the bonus.

Fifth, some metric measurements have been used for quite some time in science, medicine, pharmacy, photography, optics, the electrical industry, some ammunitions, and many international sports. The automobile industry is increasingly retooling toward metrics. Therefore, the changeover should not come as a total shock.

Too often, people tend to dismiss reasons like these as rhetoric. For many, the metric system remains a list of strange words to be memorized and strange concepts that become confused with the existing system (of which our present understanding is notoriously weak). The skeptics continue to say, "Why can't we stay with what we've got?" "To change now would be too difficult and too expensive." "It's all right for scientists to use the metric system, but, after all, very few of us are scientists." "I've tried to learn it, but it was just too tough."

Only you can decide which arguments are rhetoric and which are compelling; but your decision will ultimately depend on your own understanding of the metric system.

Encounters for Meeting the Metric System

If you have never dealt with the metric system, or if it is only a dim memory from some distant science or mathematics course, the activities which follow should prove to be significant encounters for assimilation and accommodation. The resultant learning will be meaningful if it stems from *your actions*. Don't expect to internalize the metric system simply by reading about it. For anyone—not just for children—words have only as much meaning as the experiences one brings to them. (*Experience* is derived from the Latin *experiri*, "to try out.") You, too, can learn best by doing. If you've already mastered metrics, this section will serve mainly as reinforcement with little or no adaptation required from you. However, as you examine the presentations, you might gain some insight on examples of Piagetian applications.

Length in Metrics   Examine Figure 4-1 carefully. Then cut a piece of inelastic ribbon, adding machine paper, or any similar material exactly *10 times* as long as the length shown in Figure 4-1.

Now examine the length of the material you've cut. Compare it with the length of your arm or your leg or the width of a door. You are experiencing the basic unit of length in the metric system. This unit, called a *meter* (the symbol for which is m), is the most fundamental unit of the metric system.

The metric system is based on tens; it is a decimal system. Figure 4-2, which is one-tenth (0.1) of your meter, is also divided into 10 units, each approximately the width of your little finger (or perhaps one of your fingernails). There are 10 lengths of the total Figure 4-2 in a meter. That is, there are 10 times 10 (or 100) little-finger-size units in a meter. These units are called *centimeters* (from the Latin *centum,* meaning "hundred"). A centimeter (cm) is one-hundredth (0.01) of a meter. Mark off 100 cm on your meter. Determine the length and width of this page to the nearest centimeter. What objects around you are 1 cm long or wide? 5 cm? 8 cm?

By now you've probably surmised that a centimeter can also be divided into tenths, as shown in Figure 4-3. These units are about as wide as the diameter of a conventional paper-clip wire, or almost the thickness of a dime. How many of these tiny units are in a meter? If you've figured 1000, you're

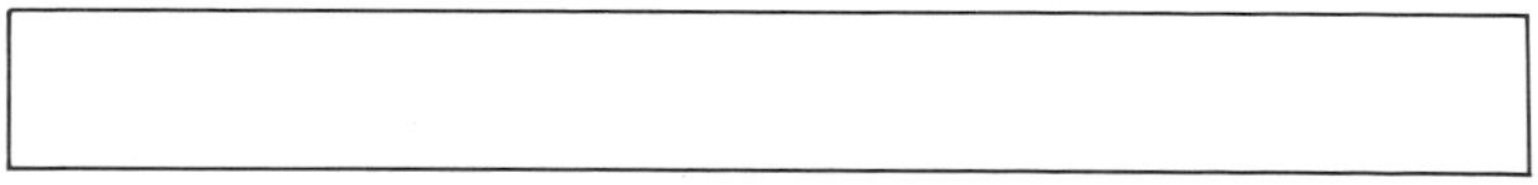

Figure 4-1

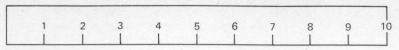

Figure 4-2

Figure 4-3

getting the idea of tens—10 tiny units in a centimeter, 100 centimeters in a meter; and so there are 1000 tiny units in a meter. These units are called *millimeters* (from the Latin *mille*, "thousand"). A millimeter (mm) is one-thousandth (0.001) of a meter. Adjust your glasses if necessary and mark off some millimeters on your meter. Measure coins, jewelry, keys, or other small objects in millimeters.

Perhaps you've made another discovery—the relationship between the metric units and our monetary system in the United States. One mill is analogous to 1 mm; one penny to 1 cm; one dollar to 1 m. Is there a metric unit comparable to a dime? Yes, but it's not used as frequently as millimeters, centimeters, and meters. That unit is one-tenth of a meter on the total length shown in Figures 4-1, 4-2, and 4-3. (See if the palm of your hand including your thumb lined up parallel with your index finger totally covers the length of Figure 4-1, 4-2, or 4-3.) The unit is called a *decimeter* (from the Latin *decimus*, "tenth"). Label the decimeters (dm) on your meter if you have enough room above your marks for the centimeters (10, 20, 30, and so on). What objects are almost as long or wide as a decimeter?

Much of the marvel of metrics is its logic and simplicity. You need only to learn a root word (like *meter*) and a few key Latin and Greek prefixes such as the following:

Milli = 0.001 as large
Centi = 0.01 as large
Deci = 0.1 as large
Kilo = 1000 times larger
Hecto = 100 times larger
Deka = 10 times larger

(*Milli*, *centi*, and *kilo* are the prefixes most commonly used.) You will not need to learn an entirely new word, then, for each different unit.

The new root words, along with the prefixes, will be more meaningful if you learn them through direct experience. For example, what reference points will you use to experience the distance of 1 kilometer (km), or 1000 m? (Since meter-length paces are rather awkward for many people, you might have to revert to a car and drive a distance of 0.6 mile.)

Mass ("Weight") in Metrics   Hold two regular-size paper clips (the kind that are 34 mm long and 8 mm wide) in the open palm of your hand. You are experiencing a mass of a tiny bit more than 1 *gram* (g). The small packages of sugar substitute usually found in restaurants also have a mass of 1 g each. Hold a nickel in your hand. You now have a mass of 5 g. At this point you might have a better understanding of a gram as a basic metric unit than you do of the concept "mass."

With the coming of both the space age and the metrification movement, both scientists and mathematicians have been attempting to correct a common error which has persisted for years. People have tended to use *weight* and *mass* interchangeably, as synonyms, or they have failed to understand the concept of "mass" at all. Mass is a distinct physical property of all objects, a measure of the amount of substance or matter that makes up an object—the amount of "stuff" of an object. Mass can also be considered as the degree of difficulty in stopping a moving object or moving a still object; in other words, it is a measure of inertia. Objects with greater mass are more difficult to move or stop than objects with less mass. The metric unit for extremely small masses is the milligram (mg), 0.001 g. Pharmacists and chemists often measure by milligrams. (Check some of your prescriptions.) Moderately small objects such as coins, pencils, keys, or candy bars can be massed in grams. Objects with greater mass, like large dictionaries, meats and vegetables, and human beings, are massed in kilograms (kg): 1 kg is 1000 g. *Weight,* however, is the gravitational force between the mass of an object and the earth. In the metric system, the unit for measuring the gravitational force is the newton (a unit not usually introduced to elementary school children).

The space age has sharpened these distinctions. If you had been on one of the lunar expeditions, your *mass* would have been the same on the earth, during the flight, and on the moon's surface, provided that your body maintained equal input and output. Your *weight,* however, would have varied considerably—from heaviest on earth to "weightlessness" while in orbit (as the earth's gravitational force was counteracted by the speed of the spacecraft), to about one-sixth of your original weight while on the moon. Although this distinction may seem trivial or overly technical, elementary science curriculum writers are increasingly impressed with its importance and are including it in their materials. (This is also true of the inclusion of metrics in general.) Therefore, you should understand the difference, so that if a teachers' guide says "Mass 15 g baking soda," you won't think it's a misprint or find it incomprehensible.

Volume in Metrics   Find a 1-pound (lb) coffee can and fill it with water up to about 2 mm from the top. Also, find a ½-gallon (gal) milk carton. Measure 11.3 cm vertically from the outside bottom and mark all four corners; then cut off the remaining top. You should have a near-cubic container. Pour the water from the 1-lb coffee can into your milk-carton "cube." (Or, if you could not locate a coffee can, fill the "cube" with water up to 1 or 2 mm below the top.) What you now have in the cube is 1 liter (L). The *liter* is the basic metric unit for volume (amount of space occupied) or capacity (holding power).

Now get a teaspoon and fill it with water. The teaspoon has a holding

Figure 4-4

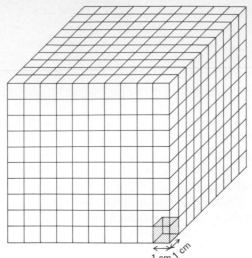

1 cm 1 cm

Figure 4-5

capacity *not* of the smallest unit of a liter but of five of these smallest units, called *milliliters*. One milliliter (ml) is only 0.001 L. How much is this, really—since it's rather difficult to envision one-fifth of a teaspoon of water?

First you'll need to construct an open cube similar to your milk-carton cube. This new cube must be 10 cm by 10 cm by 10 cm. Cut five squares each 10 cm by 10 cm from rather heavy posterboard or cardboard. Tape the sides securely together both on the insides and on the outsides, as shown in Figure 4-4. Then spread out a plastic storage bag (the 1-gal size is fine) in order to line the inside of the cube. Let the sides of the plastic bag overlap the outside of your cube. Then pour water from either your 1-lb coffee can or your milk-carton cube into your plastic-lined cube. Behold, another liter!—equivalent to a 10-cm cube, which can also be called a *cubic decimeter*. What has all this got to do with understanding a milliliter? To discover that, do two small tasks: think about the prefix *milli* and study Figure 4-5 carefully. Perhaps you can answer the question yourself. Note the dimensions of the small gray cube which appears to be inside the lower right corner of the cubic decimeter in Figure 4-5. This shape represents a *cubic centimeter* (cm^3). You would need 1000 cm^3 (or 10 cm × 10 cm × 10 cm) to fill the cubic decimeter or liter. The prefix for 1000 is *milli;* therefore 1 milliliter (ml) is equivalent to 1 cm^3. (If it's still difficult for you to envision a cubic centimeter, either mold one from clay or draw one.) If you've been in or around hospitals, you've probably heard doctors, nurses, or technicians refer to ''cc's''; cc is simply another symbol for *cubic centimeter*. Thus 1 cm^3, 1 ml, and 1 cc all have the same volume. We tend to use milliliters when dealing with liquids or powdery substances which have no shape of their own and cubic centimeters to refer to how much space an object occupies.

Temperature in Metrics   This is the final introductory encounter for meeting the metric system.

Find a Celsius thermometer—you will have to buy or borrow one if you don't have one already. (A Celsius thermometer is the same as a ''centigrade'' thermometer. The present trend is to use the term *Celsius,* in order to give credit to the developer of the centigrade scale, Anders Celsius, a Swedish astronomer of the 1700s.) Using a Celsius thermometer, practice taking some readings of the temperature of air in various locations. For example, record the temperature in a shaded or sunny spot outdoors, in the middle of a room, on a wall near the floor and near the ceiling. Record the temperature of ice water and that of boiling water. These experiences will help you understand some common Celsius references.

If a room's thermostat is set at 68° Fahrenheit (F), the Celsius (C) reading is 20°. The freezing point of water on the Celsius scale is 0°C, whereas the boiling point is 100°C. On the Fahrenheit scale, water freezes at 32°F and boils at 212°F. The difference between boiling and freezing is 180°F and 100°C. Therefore, the difference between degrees on the Celsius scale is much greater than the difference between degrees on the Fahrenheit scale—almost one to two. This is most evident when you consider normal human body temperature, which is 98.6°F and 37°C. We don't consider a "Fahrenheit fever" of 1 or 2°F very serious. However, a body temperature of 38°C is considered feverish; 39°C, very feverish; and 40°C, dangerously feverish.

Some Metric "Mechanics"   It seems as though the more precise a discipline becomes, the greater the need for a specialized language or vocabulary is. There must also be ways of abbreviating or symbolizing that language, and rules governing its correct usage. This is true of medicine, music, engineering, optics—in fact, of practically any specialized area, including the metric system. The following sections deal with rules for using metric symbols. In addition, some suggestions (not rigorous rules) will be given for implementing the metric system in doing science with children.

Symbols   There is a universally established set of symbols for the metric units. The following lists show some of the metric units, their symbols, and their numerical relationships to the base unit. (Please don't struggle to memorize the lists. They are here only to show you some patterns and some of the logic of the metric system.) An *asterisk* designates the most commonly used units.

Units of length

*1 millimeter (mm) = 0.001 meter
*1 centimeter (cm) = 0.01 meter
 1 decimeter (dm) = 0.1 meter
*1 meter (m)
 1 dekameter (dam) = 10 meters
 1 hectometer (hm) − 100 meters
*1 kilometer (km) = 1000 meters

Units of mass

 1 milligram (mg) = 0.001 gram
 1 centigram (cg) = 0.01 gram
 1 decigram (dg) = 0.1 gram

*1 gram (g)
1 dekagram (dag) = 10 grams
1 hectogram (hg) = 100 grams
*1 kilogram (kg) = 1000 grams

Units of volume

*1 milliliter (ml) = 0.001 liter = 1 cubic centimeter (cm^3 or cc)
1 centiliter (cl) = 0.01 liter
1 deciliter (dl) = 0.1 liter
*1 liter (L)
1 dekaliter (dal) = 10 liters
1 hectoliter (hl) = 100 liters
1 kiloliter (kl) = 1000 liters

Unit of temperature

*Degree Celsius (°C)

Since m, g, L, and so on are symbols rather than abbreviations, do not use periods with any of the metric symbols unless they occur at the end of a sentence. The symbols are all lowercase except L for liter, which is often capitalized when it stands alone so that it will not be confused with the numeral 1. (It is lowercase in combinations, such as ml.)

Do not capitalize the unit names other than Celsius (except, of course, at the beginning of a sentence).

Never pluralize the metric symbols: 5 grams is 5 g, not 5 gs or 5 g's.

Comparisons   Do not become overly scrupulous in memorizing (or inflicting upon your students) metric trivia such as the exact comparisons between a meter and a yard, or a gram and an ounce, or a liter and a quart. But if you have formed accurate concepts of the English or customary units, you may find it useful to use them as references for the related metric units. Depending on your understanding, the following general comparisons might be meaningful:

1 m is slightly longer than 1 yard (yd)
1 g is considerably less than 1 ounce (oz)
1 kg is slightly more than 2 lb
1 L is slightly more than 1 quart (qt)

These comparisons are not important for children, since their understanding of the English units is often unclear or incomplete. The best rule to follow is to keep the two measurement systems entirely separate.

Grade-Level Applicability

There are no conclusive data on the applicability of metric concepts at various grade levels. We have to operate mainly on the basis of Piagetian research related to measurement in general, the tentative results of England's ten-year metrification plan (which terminated in 1975), and our best hunches along with good common sense about teaching and learning. The following guidelines seem reasonable.

For the primary grades, these topics should be emphasized:

Measuring in general with any kinds of units prior to the introduction of standardized units

Introducing the meter mainly in terms of "more than," "less than," and the serial ordering of objects

Introducing the centimeter—also in terms of "more than," "less than," and serial ordering

Introducing the Celsius thermometer, which could initially be color-coded into temperature ranges (such as "cold," "warm," and "hot") rather than specific degrees

Introducing only the kilogram, *not* the gram, in upper primary grades

Introducing the liter in upper primary grades

For the intermediate and upper elementary grades, it is probably more effective to integrate metrics into the science and mathematics curricula rather than present the concepts in isolation. All that was introduced in the primary grades should be reinforced and expanded. The following topics should also be investigated and applied:

A brief history of the metric system
The milliliter
The kilometer
Decimal notation
Celsius temperature
Area measurements of square meter and square centimeter
Gram masses
Milliliters and cubic centimeters

These topics are offered as suggestions, not absolutes. Use them judiciously. Always consider the child's readiness and capability to interact both physically and mentally with concepts and principles of measurement. For example, if the meter is physically too cumbersome for small children to handle, begin with the centimeter. The same is true if a child cannot yet count to 100. *Do not move children into metrics too quickly.* (Move your colleagues or the children's parents instead.) Realize that it is fruitless to try to present metrics to a nonconserver of number and length. How meaningful will measuring with a meter be to a child who perceives that the farther away an object becomes, the more its length increases? Or to the child who believes that the centimeter ruler itself changes as it is moved along the object being measured? Measuring volume is an even more advanced concept. Many of us probably still have problems selecting on the first try the best container to hold the leftovers from dinner—a typical volume problem.

There are few concepts or skills that we grasp, master, and retain immediately. Understanding the metric system is no exception. In fact, there is nothing sacrosanct about presenting metrics that should impel one to deviate from the normal pattern of teaching and learning: moving from the concrete to the representational or pictorial to the abstract. If you hope to reach the point where you automatically "think metric," you'll need a wealth of practical experiences. The next section should provide at least a small contribution toward that goal.

Improving Your Metric Measuring Skills

If you have carried out all the preceding exercises, you have already constructed some of the materials you'll need for the coming activities. Even if you have available a good supply of commercial metric measuring instruments, you will grasp the metric system more fully if you make and use your own metric materials. At this point, the formation of meaningful concepts is far more important than the reading of precise but perhaps meaningless numerals. That old Chinese proverb, "I hear and I forget; I see and I remember; I do and I understand," is especially important when learning the metric system.

The next activities are intended to give you additional practice in using some of the metric units. More specifically, they are designed to give you more suggestions on (1) gathering and constructing metric equipment, (2) establishing more reference objects as bases for comparisons, (3) improving your estimating skills, and (4) becoming more proficient in quantifying your observations with metric standards.

These four points make up the format for most of the following activities:

A. Materials
B. Reference objects
C. Estimations
D. Actual measurements

You might find it helpful to make a record of the results for each activity. Table 4-4 is one means of organizing your information: this general format will be used in each of the following application activities.

Linear Measurements
Meters

A. Materials   Meters can be made from inelastic ribbon, adding machine paper, slats from old venetian blinds, dowel rods, wood strips cut at a lumberyard, inelastic rope, and insulated electrical wire.

B. Reference objects   It is important that you select *your* most meaningful reference objects. Investigate any of the following, for example:

Table 4-4
METRIC OBSERVATIONS

Type of metric measurements:	
Materials used:	
Reference object(s) selected:	
ESTIMATIONS	ACTUAL MEASUREMENTS
1.	1.
2.	2.
Etc.	Etc.

Distance of your hipbone from the floor as you stand erect

Height of your kitchen table

Width of a door or window

Height of a door or window

Distance from your right shoulder to the tip of your middle finger when you are holding your left arm out perpendicular to your side

Any other object or distance that approximates 1 m

Select your best, most unforgettable, reference for a meter.

C. Estimations Use your meter reference and estimate:

Height of a chair
Length and width of a table
Distance between the floor and a wall light switch
Height of the ceiling
Any additional objects or distances of your own choice

D. Actual measurements Determine the accuracy of your estimates. How well can you measure in meters?

Centimeters

A. Materials Many 12-in. rulers have centimeters designated on one side. (If you use these with children, place masking tape over the inch designations to avoid confusion.) You can also calibrate index cards, strips of tagboard, popsicle sticks, tongue depressors, or inelastic lengths of ribbons in centimeters.

B. Reference objects   Compare the following with a centimeter:

Widths of your fingernails
Width of a popsicle stick
Jewelry you usually wear
Height of lines on notebook paper
Any other objects which approximate 1 cm

Select your best reference for a centimeter.

C. Estimations   Use your centimeter reference and estimate the length and width of:

Leaf
Stick of chewing gum
Pen or pencil
Shoe
Any other objects of your choice

D. Actual measurements   Check the accuracy of your centimeter estimations. Were your initial attempts at estimating less or more accurate than your later attempts?

Millimeters

A. Materials   Calibrate any of your centimeter materials to millimeters. Refer to Figure 4-3.

B. Reference objects   Compare the following with a millimeter:

Diameter of thin lead from a mechanical pencil
Thickness of a dime
Width of a straight pin or safety pin
Thickness of a small rubber band
Thickness of any of the lines of your palm
Other objects of your choice which approximate 1 mm

Select your best reference for a millimeter.

C. Estimations   Use your millimeter reference and estimate:

Width of a blade of grass
Length, width, or thickness of any jewelry you usually wear
Thickness of any keys you have

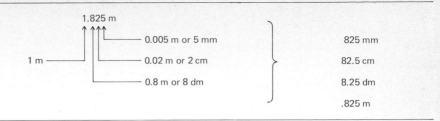

Figure 4-6 Placement values.

Thickness of a nickel
Any other objects of your choice

D. Actual measurements Test the accuracy of your millimeter estimations. Do you tend to overestimate or underestimate? If necessary, select some additional objects to measure in meters, centimeters, and millimeters.

Linear Combinations Suppose that you saw a person's height recorded as 1.825 m. Would you suspect the person to have the potential for being a jockey or a basketball player? What does 1.825 m really mean? Consider the place values shown in Figure 4-6. Then actually measure and mark this hypothetical person's height (1.825 m) on a wall. Identify each component: 1 m, plus 8 dm, plus 2 cm, plus 5 mm. Many metric workbooks call for the mental gymnastics of transposing a numeral into different units. For example, converting 1.825 m into 18.25 dm or 182.5 cm or 1825 mm is relatively simple once you discover the "magic" of metrics: decimal-point movements. However, it is not as easy to conceptualize the results of this logic. It takes quite some effort to fully perceive the equivalence of 182.5 cm, 1.825 m, 1825 mm and 18.25 dm. These types of conversions might be interesting for adults, but they are inappropriate for elementary school children. They present too much potential for false accommodation. It is far more important that children learn the concepts of meter, centimeter, and millimeter and that they learn when to use the correct units—for example, meters for large objects or distances, centimeters for smaller objects or distances, and millimeters for very small objects.

Mass Measurements

Grams

Preparation You'll need a balance and some gram masses for the next activities. Even if you have access to these materials now, you may eventually find yourself in an instructional situation where such supplies are not available or are available only in small amounts. Therefore, it would be wise to construct your own materials and build up a reservoir of models which some of your students might eventually construct. Examine the examples in Figure 4-7 and decide which balance or balances to construct or assemble. As you make your balance, keep in mind the following basics:

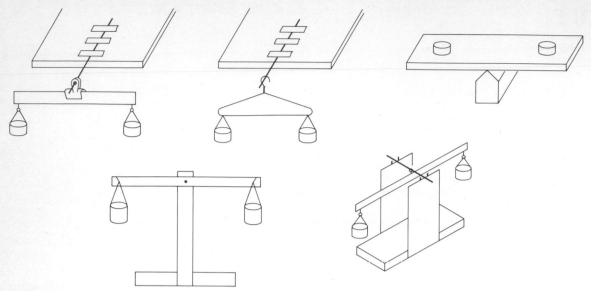

Figure 4-7 Examples of homemade balances.

1. Have a firm base or support system so that the apparatus will not topple over.
2. Try to eliminate as much friction as possible at the fulcrum, the point where the center of the balance beam pivots.
3. Use rather heavy, strong material so that the balance will not be too sensitive, will not constantly need to be readjusted after each measurement, and will not fall apart when greater masses are measured.
4. Since it is often difficult to keep all factors equal on each side of the balance beam, attach small pieces of clay to make the necessary adjustments.
5. Test your finished product. For example, compare several pairs of objects that should have equal masses—like two identical packs of gum, two coins of equal value, or two cans of the same product. If your results are inconsistent, either make the necessary adjustments or select another design and begin again, making a better balance. Once you are satisfied with the balance you have constructed, use it for measuring some gram masses.

A. Materials The most readily available and reasonably accurate objects which can be used as references for gram masses are pennies and nickels. A penny has the mass of 3 g; a nickel, 5 g. You can use nickels as standards to make a collection of additional masses, such as dry sand or pebbles in small plastic sandwich bags. Tie or tape the bags shut if you wish to keep the contents intact. (In case you want to be exact, the plastic sandwich bag has the mass of about 1 g.) Decide how extensive a collection of gram masses you wish to make—the greater the variety, the better.

102

B. Reference objects Use your balance or balances and the set of masses you have calibrated or accumulated to measure the number of grams in the following:

Wristwatch (or any other personal jewelry)
Pen or pencil
Keys
Teaspoon
Any other objects of your choice

C. Estimations   On the basis of the gram references you have established, estimate the masses of the following:

Comb
Egg (Is there a difference between hard-boiled, soft-boiled, and raw eggs?)
Apple, orange, or any kind of fruit you usually eat
Fingernail clipper, or nail scissors
Any other objects of your choice

D. Actual measurements   Check your gram estimations. Don't be overly concerned with pinpoint accuracy. Anyone who can estimate a mass within 1 g (0.035 oz), or even within several grams, must have extraordinary sensory perception. Developing familiarity with measuring amounts of materials in grams is more important than being absolutely precise.

Kilograms   If your balance is fairly durable, and if you can collect or calibrate enough masses to amount to 1 kg (1000 g, which is about 2.2 lb), spend some time massing larger or perceptually heavier objects than those just suggested. (Try also to include a kilogram in your collection of masses.) Since many food companies are including grams along with ounce and pound designations on their products, it's not too difficult to collect items amounting to a kilogram. Then follow the format of the previous activities. Use your improvised kilogram to establish some common reference objects which you can use as bases for estimating other masses. Then check the accuracy of your estimations by trying these kilogram and gram masses on your balance.

Don't be surprised if you find your estimates faulty. You've lived for many years using ounces and pounds; yet how accurate are you at estimating amounts of produce at the grocery store? Remember that you're learning a new concept—and concept formation usually requires considerable time and many encounters or practice sessions.

Volume Measurements

Milliliters and Liters

A. Materials   If you have any metrically graduated cylinders, beakers, or flasks available, use these to calibrate your own volume-measuring devices. Tall, slender olive jars can easily be made into graduated cylinders. Try to find other

104

SCIENCE FOR CHILDREN

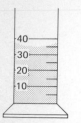

Figure 4-8 Measuring liquid in a graduated cylinder.

containers which have fairly straight sides, so that it will be easier to interpolate (estimate between units) or extrapolate (estimate beyond the last unit). If you use a graduated cylinder to measure milliliters of water, be sure to practice good laboratory techniques. Most likely, the water surface will be slightly curved, because of the adhesive quality of water. This curved surface is called the *meniscus.* When measuring volume, always hold the graduate level or place it on a level surface and keep your eyes even with the meniscus. The most accurate reading is based on the lowest point of the meniscus; for example, 35 ml in Figure 4-8. If you have no metric standards available for measuring volume, use some cooking measures instead: the information in Table 4-5 can be used to establish some basic starting points: (Use this table of equivalents *only* if you have no metric volume-measuring equipment.) You can use the 1-lb coffee can, the milk-carton "cube," or the plastic-lined cubic decimeter (see page 94) as your liter standard.

B. Reference objects   Use your calibrated milliliter measures to determine the volume or holding capacity of the following:

Your favorite coffee cup or drinking glass
Saucer
Cereal or soup bowl
Saucepan
Any other common containers you usually use

C. Estimations   On the basis of your milliliter references, estimate the capacity of the following:

10-, 12-, or 16-oz beverage bottle (fluid measure)
Juice glass
Small skillet
Salad bowl
Any other objects of your choice

Table 4-5
COMMON EQUIVALENTS

KITCHEN UNIT	APPROXIMATE METRIC VOLUME
½ teaspoon (tsp)	2.5 ml
1 tsp	5 ml
1 tablespoon (tbsp)	15 ml
⅓ cup	60 ml
⅓ cup	80 ml
½ cup	120 ml
⅔ cup	160 ml
¾ cup	180 ml
1 cup	240 ml
2 cups	480 ml
1 qt	960 ml (slightly less than 1000 ml or 1 L)

D. Actual measurements Determine the accuracy of your estimations. How does your ability to estimate and measure capacity compare with your ability to estimate and measure mass? Again, don't be overly concerned with absolute accuracy. Simply try to keep improving your estimating and measuring skills. You might eventually be able to store those leftovers in the smallest container on the first try; but there's still the problem of storing the containers in the refrigerator. We need to investigate how to determine volume in terms of how much space an object occupies.

Cubic Centimeters You can always use the formula *height × width × depth* to determine the volume of a regular cubic or rectangular object. For example, a container 9 cm × 9 cm × 7 cm (567 cm³) would be ideal for storing 500 ml or 0.5 L of soup. But do you have any idea how much space a carrot or a potato or any irregularly shaped object occupies? To determine this, the easiest method is water displacement.

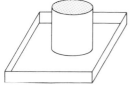

Figure 4-9

A. Materials You'll need the 1-lb coffee can again, a cake pan, and your calibrated volume measures. Remember that 1 L of water does not totally fill the coffee can; the water level is about 1 or 2 mm from the top. But for this activity the coffee can should be placed in the cake pan and completely filled with water, as in Figure 4-9. (If you do not observe how the water "heaps" higher than the top edge, you've probably not filled the can with enough water.)

B. Reference objects The amount of space an object occupies can be determined by measuring the amount of water it displaces when it is *gently* released into the can of water. Why is this so?—because two objects cannot occupy the same space at the same time. The object being measured must sink in order for this method to work accurately. You'll need to be careful that your fingers do not touch the water as you release the object. Gently remove the coffee can and empty the spilled water from the cake pan into your calibrated volume measurer. The measurement indicates the amount of space the object occupies. Now use this water displacement method with the following:

Golf ball
Your fist (lowered to the main wrinkle of your wrist)
Any type of fruit which will sink
Can of soup or vegetables
Any other sinkable objects

Estimations Use the data obtained from the reference objects to estimate the water displacement of the following:

Egg
Your hand with fingers fairly open (up to the wrist wrinkle)
Any type of vegetable which will sink
Can of food (different mass from the can used previously)
Any other objects of your choice

D. Actual measurements Test your estimations. Which results were most surprising? In what ways do the concepts of milliliters and cubic centimeters make more sense to you now?

Putting the Parts Together

Interrelationship of Length, Mass, and Volume So far, linear, mass, and volume metric measurements have been considered separately. However, there is an interplay, a unique interrelationship, among metric units that is totally absent in the English customary system. Discover how the meter, liter, and kilogram are related by following these steps.

1. First, determine if your balance is sturdy enough to handle slightly more than 1 kg on each side of the balance beam. If your present balance is inadequate, either procure or construct a stronger balance.
2. Next place your empty plastic-lined cubic decimeter on one side of the balance beam and balance it with the necessary amount of gram masses.
3. Then slowly fill the plastic-lined cubic decimeter with 1 L of water.
4. Determine the amount of mass needed to balance the 1 L of water. (Disregard the previous masses which balanced the empty cubic decimeter.)

If you've been fairly accurate, the results should indicate that 1 L of *water* (but not *every* liquid) has the mass of 1 kg. Now think about the mass of 1 ml or 1 cc (1 cm³) of water. (There are 1000 ml or 1000 cm³ in 1 L; 1 ml or 1 cm³ is 0.001 of a liter; therefore . . .) The mass of 1 ml or 1 cm³ is 1 g.

These phenomena demonstrate another of the marvels of metrics: there is, indeed, a logical relationship among linear, mass, and volume units in the metric system.

Size Yourself Up The final metric activity is illustrated in Figure 4-10. This activity enables you to give the metric system some personal meaning. After you finish collecting the data, determine how many relationships you can find. For example, find the relationship between:

Your outstretched arms and your height
Length of your eyebrow and your nose (or ear)
Circumference of your neck and your knee
Volume of your hand and your foot

How many different relationships can you discover?

If you have access to a bathroom scale calibrated in kilograms, determine your metric mass. If no such scale is available, a very close approximation of your mass in kilograms can be calculated by multiplying your mass in pounds by 0.45. You'll probably be pleased by this final metric discovery: learning your mass in kilograms may reconcile you to your waist circumference in centimeters.

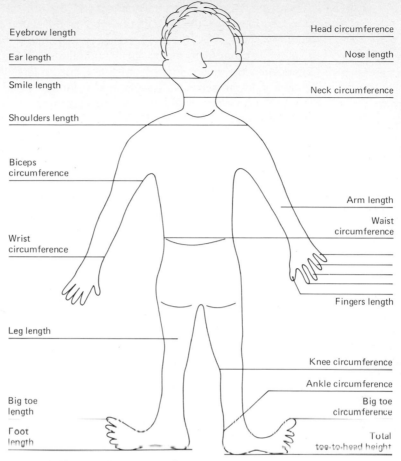

Figure 4-10   How do you size up?

The Key for Getting Started

Science begins with observing. In fact, all learning ultimately begins with input from the senses. If, in the past, you've been limiting your observations to receiving input only through the sense of sight, you've been missing 80 percent of the potential experiences. I hope the activities in this chapter have increased your capacity for absorption, your ability to take in as well as act upon stimuli from the natural world. The better you are at observing and measuring, the greater your chances for understanding as well as enjoying the objects and events around you will be, the greater your responsiveness and personal growth will be. What's more, your ability to foster children's responsiveness can also be greatly enhanced.

The degree of disequilibrium you may have experienced or may still be experiencing depends on several factors—for example, the degree of newness or strangeness of the concepts introduced, the amount of your active *doing* (as opposed to merely reading), the number of times you practiced your observing and measuring skills, and of course your attitude toward the experiences. (Certainly, other factors may also have influenced you.) However, if you have interiorized the major ideas from these encounters, the whole could well be greater than the sum of the parts. For instance, one of the most significant and far-reaching results could be the development of greater sensitivity toward children's learning in general. It's often been said that students who value yet struggle with learning become the best teachers, since they better appreciate their own students' struggles. This may or may not be true. But it is true that adults tend to forget how little they knew as children. It's to be hoped that when you present observing and measuring encounters (or any learning encounters) to your students, you will be more conscious of your own learning experiences. Your increased self-awareness and self-acceptance can help improve students' self-concepts. Your being "tuned in" to the total environment will encourage them to become better observers, both qualitatively and quantitatively.

What, ultimately, do you need to get started with science for children? Often a key is needed—and in this case the key is the sense of caring.

Encounters

1. One way to develop more awareness of yourself, of others, and of the environment is to participate in a "trust walk." Blindfold yourself and let another person lead you around to a specific spot. Remain there for a while to assimilate the environment and accommodate yourself to it. Come to know that environment. Then ask your guide to lead you away. Remove your blindfold and see if you can relocate the same spot you previously absorbed. Next, reverse roles and repeat the procedure; but select a different spot. Discuss your feelings and impressions, the use of your senses, and your general reactions.

2. Select a favorite outdoor spot near where you live. Spend some time observing this spot on a regular basis. Keep a log of your observations. What changes did you observe—both in your observing skills and in what you had been observing?

3. Develop one or more activities which will enhance the skill of observing. If possible, try out your ideas with a child or a small group of children. Revise your activity or activities if necessary. Share your results with your colleagues.

4. If you have a bent toward history, research the historical development of metrics, especially in the United States. For example, find out which Presidents seriously considered metrics for the United States and what happened to the Metric Conversion Act of 1972.

5. Develop some additional metric activities for children. Test your ideas, evaluate the results, and share these activities with your colleagues.

6. "Metrify" your living quarters or classroom. Determine and label the linear dimensions, mass, and volume of the objects you use most often.
7. Your choice: what other questions or areas of interest related to this chapter do you wish to investigate?

References

American Association for the Advancement of Science: *Science—A Process Approach, Commentary for Teachers,* 3d experimental ed., Washington, D.C., 1968.

Van Matre, Steve: *Acclimatizing,* American Camping Association, Martinsville, Ind., 1974.

Suggested Readings

On Increasing Observing Skills and Environmental Awareness

Busch, Phyllis S.: *The Urban Environment, Kindergarten to Third Grade,* Ferguson, Chicago, Ill., 1975. (This book contains over sixty lessons which are centered on four conceptual schemes: heredity-environment, interdependencies, change, and human behaviors. Both science and social studies concepts are clearly integrated.)

Dillard, Annie: *Pilgrim at Tinker Creek,* Bantam Books, New York, 1974. (The author literally lives her environmental awareness as she describes life near a creek in the Blue Ridge region of Virginia. Her descriptions and insights can greatly improve your own observing, especially for item 2 in the Encounter for this chapter.)

Headstrom, Richard: *Nature in Miniature,* Knopf, New York, 1968. (This is a resource guide for discovering the less obvious wonders of nature with the aid of a hand lens. It provides a mini trail around the world with ninety-nine scenic points of interest.)

Leopold, Aldo: *A Sand County Almanac,* Ballantine Books, New York, 1971. (The author was a conservationist long before the environmental movement became popularized. Seasonal characteristics of nature and regional differences are described. The last section, on environmental ethics, is invaluable reading.)

McInnis, Noel: *You Are an Environment: Teaching/Learning Environmental Attitudes,* Center for Curriculum Design, Evanston, Ill., 1972. (McInnis's definition of *environment* as anything which is influencing or being influenced by something else will cause you to take a much broader view of self-awareness and environmental awareness.)

Russell, Helen Ross: *Ten-Minute Field Trips,* Ferguson, Chicago, Ill., 1973. (Very practical, easy-to-follow suggestions are given for using the school grounds or immediate area for environmental studies. One of the introductory sections, "Of the Value of Saying, 'I Don't Know,'" will give you courage. The major sections give a wealth of information and ideas for activities.)

Van Matre, Steve: *Acclimatization,* American Camping Association, Martinsville, Ind., 1972. (This book is one of the most popular primers on developing a sensory and conceptual approach to environmental awareness and involvement. It was the foundation on which Van Matre's *Acclimatizing* was built.)

Watts, May Theilgaard: *Reading the Landscape,* Macmillan, New York, 1964. (The author shows how to read one's environment by observing natural signs rather than printed or painted words. A vast amount of content can be subtly assimilated along with more astute observing skills.)

On Metrics

Bitter, Gary G., Jerald L. Mikesell, and Kathryn Maurdeff: *Activities Handbook for Teaching the Metric System,* Allyn and Bacon, Boston, 1976. (In comparison with most books on metrics, this is a tome. The background information, activities, history, and summary tables are thorough but not overwhelming. This is an excellent resource book.)

Henry, Boyd: *Teaching the Metric System,* Weber Costello, Chicago, Ill., 1974. (In only forty-six pages, a great deal of background information and numerous classroom activities are presented. This is a very practical booklet.)

Higgins, Jon L. (ed.): *A Metric Handbook for Teachers,* National Council of Teachers of Mathematics, Reston, Va., 1974. (This book contains seventeen important articles from a wide variety of authors on measurement concepts and ideas for introducing and teaching the metric system.)

Leffin, Walter W.: *Going Metric,* National Council of Teachers of Mathematics, Reston, Va., 1975. (This is another short but very informative book— forty-six pages. More emphasis is placed on the development of measurement standards, especially in the United States, and on some pedagogical guidelines than on instructional activities.)

Ostergard, Susan, Evelyn Silvia, and Brandon Wheeler: *The Metric World: A Survival Guide,* West, New York, 1974. (The authors present a vast amount of information, with ample opportunities for practice and applications.)

CHAPTER 5
BUILDING THE BASIC FRAMEWORK

111

The Importance of the Process Skills

The teaching of science to children has never become as prescriptive as, for example, the teaching of reading and mathematics. The leading professional organizations for science teachers have never officially endorsed and promulgated a formal statement describing and prescribing specific facts, principles, or topics to be presented at each age or grade level. (However, some sources for guidelines and suggestions from organizations of science teachers can be found in the Suggested Readings.) Probably one of the main reasons for this seeming lack of direction is the legitimate concern that specification would stifle the real spirit of science. There is a real danger that the acquisition of facts or principles would become an end in itself, and the dual nature of science—as product and process—would not be understood.

Instead of building a science program on the shifting sands of content, many science educators of the post-Sputnik era have tended to consider process skills, or mental operations, as a stronger foundation. These mental operations best represent the spirit of science and the intellectual tools that scientists use. But more important, they are skills of thinking that we can all use for achieving more rational thought and action. We have already considered two of these skills—observing and measuring.

It would be presumptuous to attempt to identify all the thinking skills. But we can try to envision how a thoughtful person might handle a situation or problem. This person may perform any or many of the following mental operations: comparing, describing, classifying, using numbers, graphing, distinguishing space-time relationships or symmetry, ordering or seriating, analyzing, synthesizing, building models, inferring, predicting, hypothesizing, collecting and interpreting data, identifying and controlling variables, defining operationally, testing and validating, communicating results. Many or all of these actions can be combined in the most cumulative of skills—namely, experimenting.

Why are these processes so important for science and so basic to all thinking and learning? You will probably develop your own answers to this question after you have become more familiar with some of these mental skills, three of which—distinguishing spatial and temporal relationships, classifying, and communicating—are focused on in this chapter. But some answers can be considered now, and you may want to reconsider them later.

First, learning occurs within the learner as a result of physically or mentally responding to a learning encounter. There are countless ways to respond to a learning encounter. But in the final analysis all possible actions are synonymous with the process skills. These are the mechanisms which trigger assimilation and accommodation—that is, the essential components of real learning.

Second, the process skills equip the learner with the tools for making sense out of chaos, order out of disorder. They are not fads or gimmicks; their applicability continues indefinitely. "Facts" may become outdated in a short period of time; but the process skills enable the learner to make continual discoveries.

Third, the process skills arouse the sometimes somnolent sense of wonder. They entice one to activity rather than passivity. They lead to new awakenings, explorations, and revelations.

Fourth, when the process skills become freely selected patterns of behavior, they serve to regulate the disequilibrium-equilibrium cycle of learning. Using them enables the learner to move from the "I don't get it" state of being bothered to the "I've got it" state of learning. The more often they are used successfully, the more likely it becomes that they will be chosen for use in a new situation. Thus they regulate learning and motivate the learner.

To sum up: essentially, the process skills are means toward greater rationality—not ends in themselves. They are applicable in *all* areas of learning, not just science. And, like children's spontaneous curiosity, they are "naturals" for elementary science.

If you believe that the primary purpose of elementary science is to facilitate the development of children's thinking, then you will conclude that the process skills provide the framework for this development. Perhaps an analogy might help: Compare the process skills to the skeletal system of the human body. Observing and measuring are the backbone or vertebral column and serve as the support system. The basic skills to be considered in this chapter—distinguishing spatial and temporal relationships, classifying, and communicating—are like the ribs. They are closest to the heart and breath of scientific endeavors. The skills which will be presented in Chapter 6 (inferring, predicting, and hypothesizing) and Chapter 7 (controlling variables, interpreting data, and defining operationally) are similar to the appendages. They enable the learner to reach out and move forward. Experimenting, the skill which incorporates the most logical operations, is like the skull, which houses the brain. To continue the analogy: If the backbone is weak, the rest of the skeletal system is affected. If the entire framework is poorly developed or unhealthy, the entire body (as well as the whole person) will be impaired.

These preliminary comments will probably be meaningless rhetoric to you unless you experience the processes, learn to use them with children, and come to value them. In this chapter and in Chapters 6 and 7, several activities will be presented for each process. Generally, the activities will be sequenced from least difficult to most difficult. Although they have been developed for your use, they can easily be modified for use with children. Some of the ideas are derived from commercial elementary science programs which will be described in greater detail in Chapter 9. The format used for some of the activities in Chapter 4 will also be used here: (A) introduction, (B) materials, (C) procedures, and (D) extensions.

One final suggestion, or warning, before you begin these activities: It is virtually impossible to present each process as a discrete, isolated, totally unique action. For example, observing cannot really be separated from any of the other skills. Often, you will be using several skills in order to better understand one particular skill. You will soon see that the overlap and interplay among the processes is not only inevitable but also essential.

Distinguishing Spatial and Temporal Relationships

The skill of distinguishing spatial and temporal relationships is most akin to the skill of observing. The major actions performed in observing involve the use of the senses and measuring. Few objects are ever observed in total isolation. Objects generally exist, or are situated spatially, with other objects. They also are observed differently according to the position of the viewer. If four people, for example, were viewing a mountain—one from a helicopter, one from the base of the mountain on the west side, one from the base on the east side, and one from midway up the steepest slope—their views would be quite different. An object, or its relationship to its surroundings, can also change over a period of time.

In order to describe the physical environment better, you will need to sharpen the skill of distinguishing spatial and temporal relationships.

Activities

Space and Shape

A. Introduction   The ability to identify and distinguish basic shapes is of great importance for a young child's perceptual development. It is also an important component of aesthetics or simply being aware of and more comfortable with one's environment. What are the predominant shapes in the space of your environment?

B. Materials   A "scorecard" like the one shown in Table 5-1.

C. Procedures
1.   Select a room (either at home or at school) and analyze the variety of shapes in the space of the room or in one section of it. Look for shapes corresponding to those suggested in Table 5-1; also, add your own choice or choices.
2.   Place tallies under the appropriate categories to represent the number of different shapes present. Which shapes are leading?
3.   Next, select an interesting space outdoors and repeat procedures 1 and 2.
4.   Compare your outdoor results with your indoor results.
5.   Which objects were made up of the most combinations of shapes?
6.   What changes might occur in your results after several hours? After several days? Weeks? Months? Years?

D. Extensions   A "shape scavenger hunt" would be appropriate for younger children. Have them begin with a simple task like finding one shape at a time: then have them find combinations of two shapes; and finally, offer a grand prize

Table 5-1
SPACE AND SHAPE SCORE CARD

Square ☐		Rectangle ▭		Triangle △		Your choice	
(Inside)	(Outside)	(Inside)	(Outside)	(Inside)	(Outside)	(Inside)	(Outside)
Rhomboid ◇		Circle ○		Ellipse ⬭		Your choice	
(Inside)	(Outside)	(Inside)	(Outside)	(Inside)	(Outside)	(Inside)	(Outside)

for the object with the greatest number of identifiable shapes. If you notice that some children have difficulty concentrating on one limited area, give them viewing tubes like empty paper-towel rolls or large index cards rolled into tubes. Ask them to look at one spot through the tube then slowly move the tube to an adjacent spot.

One of the major uses of the skill of distinguishing spatial relationships is in storing three-dimensional objects. Test your own skill with this extension: Select a fairly large box—one at least 4 dm × 3 dm × 2 dm. Then collect a wide variety of ten to twenty much smaller boxes that have secured lids. Determine how many smaller boxes can be fitted into the large box. Two rules must be followed: no small boxes may be placed inside medium-size boxes, and no boxes may extend over the top of the large box. Once you've solved the storage problem, make a crude sketch of your arrangements. Then determine if someone else can do a better job—perhaps even in less time. Be prepared to challenge your nearest competitor. Think about this activity the next time you pack or move or rearrange storage space.

Halves and Halve-Nots

A. Introduction Examine Figure 5-1 (page 116). In what ways are the drawings in set A alike? How are the drawings in set B similar? What does set A have that set B does not have?

Figure 5-1 Halves and halve-nots.

Determine how each of the drawings in set A could be folded so that one half is exactly like the other half. Could the drawings in set B also be folded into identical halves? Regardless of how many ways you might try to fold the figures in set B, you will not come up with matching halves. The figures in set A, which can form identical halves when folded, have the property of *symmetry*. Symmetry can be demonstrated not only by folding but also by drawing a real or imaginary line which divides the object into two identical halves. This line—folded, drawn, or simply imagined—is called the *line of symmetry*.

Some three-dimensional objects also possess the quality of symmetry. For instance, if you were to slice an apple or orange exactly in half and then look at the two exposed surfaces or outlines, you would see another example of symmetry. The slicing knife has created a *plane of symmetry*.

This activity will give you an opportunity to discover the halves and the halve-nots of symmetry in your environment. (The activity can be performed with a partner.)

B. Materials   (1) The contents of your junk drawer, tool chest, purse, or glove compartment, or *any* random collection of odds and ends; (2) some plain paper; and (3) scissors.

C. Procedures

1. Examine your collection of assorted objects.
2. Before separating them according to which have symmetry and which do not, decide how exact or precise you want to be. For example, do you want to consider only the general shape or composition of each object? Or do you want to include even the slightest details—like printed words, chipped paint, nicks, or scratches? (Children occasionally become very fussy over these matters, and so the rules should be clearly established in advance.)
3. Then determine which objects have the property of symmetry and which do not. What were the bases for your choices?
4. Look for additional examples of symmetry around you—both indoors and outdoors.
5. Draw as exactly as you can, and then cut out, three shapes: a rectangle, a square, and a circle. Determine all the possible lines of symmetry each shape has. What objects have two lines of symmetry? Three lines of symmetry? Four lines of symmetry? An infinite number of lines of symmetry?
6. Now reexamine your collection of symmetrical objects. Which of them have more than one line of symmetry?

D. Extensions There are different types of symmetry. The kind you have been considering by folding or identifying equal halves is called *bilateral* ("two-sided") symmetry. While examining the cutout circle, you might have discovered that it could be turned or rotated and still look the same. This is an example of *rotational* (or *radial*) symmetry. An equilateral triangle or a square, when turned from one edge to the next edge, is another example of rotational symmetry. You may also have noticed a third type of symmetry present in the arrangements of bricks in a wall, the placement of leaves or buds on branches in either opposite or alternate positions, the repetitious patterns of designs in material or carpeting, or the regularity of venetian blinds or tiles. This type of symmetry is called *repeating patterns* or *repeating symmetry.* Identify or cut out examples for each of the three types of symmetry. Which objects have more than one type of symmetry?

This next extension utilizes both bilateral symmetry and repeating patterns. (With a bit of ingenuity it could also incorporate rotational symmetry.) Fold a square or rectangular piece of paper exactly in half and cut out any sort of shape or design from the folded edge. (An example of a single-fold cutout is shown in Figure 5-2a on page 118.) Do *not* unfold the paper yet. First try to draw what you think the design will look like when the paper is unfolded. After you've made your guess, unfold the paper and check its accuracy.

If you like challenges, try different varieties of designs and even double folds, as suggested in Figure 5-2b. That is, fold the paper in half as in the preceding step, then fold it again and cut out designs along both folded sides.

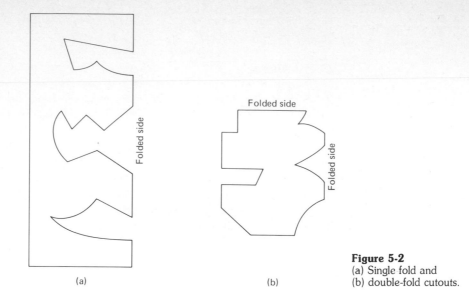

Figure 5-2
(a) Single fold and
(b) double-fold cutouts.

What spatial relationships did you discover? What discoveries did you make about your own ability to distinguish spatial relationships?

Mini-Explosions

A. Introduction   It is not always easy to be conscious of objects in terms of space *and* time relationships. We miss many of the changes occurring around us because they happen either too quickly or too slowly. The lighting of a match, the bursting of a balloon, the explosive landing of a drop of rain onto a leaf are events which can be captured only by special photography. The same is true of the germination of a sprout as it finally cracks through the seed coat, the emergence of a moth from its cocoon, or the sometimes subtle and sometimes dramatic changes in the landscape after a rain. There are probably numerous changes occurring around you right now. Which ones can you detect?

In this activity you will be able to observe relatively fast, medium, and slow changes. The challenge is to focus on the spatial and temporal relationships simultaneously—as if your eyes were the lens of a movie camera. (Because it is necessary to time the events, do this activity with one or two partners.)

B. Materials   (1) Two identical tall clear glasses (the taller the better); (2) a plastic container of food coloring that has a built-in dropper (otherwise, use a medicine dropper with food coloring); (3) a source of very hot water; (4) a source of very cold water (the greater the temperature extremes, the better); and (5) a watch or clock with a second hand.

C. Procedures
1.   Place the two glasses where it is comfortable for you to view them at eye level. Fill one with very hot water and the other with an equal amount of

Table 5-2
OBSERVATIONS OF MINI-EXPLOSIONS

	HOT WATER	COLD WATER
INITIAL REACTIONS		
AFTER 15 SECONDS		
AFTER 30 SECONDS		
AFTER 45 SECONDS		
AFTER 60 SECONDS		
AFTER 2 MINUTES		
AFTER 3 MINUTES		

very cold water. Wait for the water in both glasses to become still. (If you look from beneath the water level, you can usually detect if any tiny air bubbles are still moving.)

2. While you are waiting for the water movement to decrease, prepare both physically and mentally for the next steps. You'll be dropping a small amount of food coloring into the glass of hot water first and observing the results. After several minutes, repeat the same procedure with the cold water—that is, try to drop in the same amount of food coloring from the same height. As you observe the events, also be conscious of the spatial relationships between the food coloring and the water during brief segments of time.

3. Describe the mini-explosion that occurred when the food coloring was dropped into the hot water. Table 5-2 shows one way of organizing your observations. After several minutes, repeat the procedures and record your observations of the food coloring dropped into the cold water.

4. How do the results differ? How are they similar? Which time segment or reaction was most difficult to describe? Summarize the spatial and temporal relationships which you observed.

D. Extensions Young children especially need considerable practice in recognizing fast and slow changes, regular and irregular occurrences. Such awareness is prerequisite for the sometimes overwhelming task of learning to "tell time." The preoperational child especially has difficulty in recognizing and sequencing spatial and temporal events. Simple tasks like observing a melting ice cube or butter patty can be helpful—especially if children draw the events and show how one form disappears and another appears. You should realize that if changes occur quite slowly, the young child may lose track of the events and even of the identity of the object or organism which has changed. For example, some children are totally unaware that the plant they grew was once the sprout or seed they planted—or that the beetle they now watch was that same mealworm that crawled across their fingers. Parents should be encouraged to make pictorial time lines for their young children, displaying a child's growth and development over a period of time.

A simple sequence of four or more cartoons, preferably with no words, can also provide practice in recognizing spatial and temporal relationships. Cut apart the series and mount each picture on an index card for easier handling. Then ask the child to arrange the pictures in some order of events and to explain the reasons for the sequence. (This is a good way to combine science and language arts.) Determine the child's awareness of space and time relationships as clues, as opposed to random guessing or meaningless ordering.

Rates of sedimentation also include fast, medium, and slow changes. Place several handfuls of soil in a clear tall jar. Fill the jar almost completely with water, tighten the lid, and shake the contents for about 1 minute. Place the jar where it cannot be disturbed and observe the results over certain periods of time—for example, for several minutes, after several hours, then daily for a week. Try using a variety of soil samples. What changes and space-time relationships can you detect? Similar activities could include mixtures of colored water, alcohol, and cooking oil; or water, fine sand, cornmeal, and talcum powder.

Log Puzzles

A. Introduction   Do you sometimes feel that life itself is a puzzle in space and time? What will the final product look like? How do all the bits and pieces fit together? And when is there ever enough time to . . . ? It's difficult to formulate a whole picture, either physically or mentally, without having a clear view or understanding of the related components. This activity will provide you with an opportunity to practice your puzzle-solving skills, and perhaps even your patience and perseverance.

B. Materials   (1) Ten or more strong rubber bands; (2) a quartered or halved fireplace log; and (3) access to a good electric table saw or a cooperative owner of a good electric saw. The log should be quite dry and well weathered, especially at the two ends. The greater the variety of distinguishable characteristics, the better—bark on one side, for example, or different colorations, and textures, or insect borings. Saw the log (or have it sawed) into six to twelve pieces of varying lengths and thicknesses, and with various angles. Figure 5-3 shows some unassembled and assembled log puzzles.

C. Procedures
1. Randomly place the log puzzle pieces in an uncluttered work area. Spend some time observing all the physical properties of each piece. Speculate on possible relationships between pieces.
2. *Caution*: *Before* you try to put the log back together, decide on the reason or reasons for each choice or move. Analyze the fit in terms of the matching physical properties and, most important, the spatial relationships.
3. Then check the accuracy of the reasons for your choices by trying to put the log back together in as few moves as possible. Use the rubber bands to secure the pieces if they keep slipping apart.
4. After you have reassembled the pieces into the original log shape, consider

(1)

(2)

(3)

(4)

Figure 5-3 Log puzzles.
(Photograph by
Elizabeth Craft.)

how time relationships are involved in the activity. (The weathering process is a temporal factor. The weathered surfaces will always be the outer surfaces of the pieces. The newly exposed surfaces, those that were most recently sawed, will generally have a lighter color than—or certainly a different color from—the weathered surfaces. The "growth rings" present and the orientation of the wood grain also have to do with time.)

5. How sharp are your log-puzzle skills? What is the correlation between your patience and your ability to distinguish spatial and temporal relationships?

D. Extensions Use fewer and less bulky pieces for younger children. Also, try to use logs that have more distinguishable characteristics. If you use this activity with children, watch for their tendency to try to force matches that simply do not fit and their ability to use reversibility in their thinking and actions. You can gain some insights into their problem-solving approaches and their degrees of egocentricity.

There are a number of variations which make this activity more game-like.

For instance, you or the children could develop a set of rules, such as: (1) No piece may be moved until one tries to make a match. (2) One point is scored for each correct choice. (3) One point is deducted for each incorrect choice. (4) The person or team scoring the most points in the least time wins.

The activity could be made even more challenging by secretly removing one of the pieces. Explain that a piece is missing and ask the children to use modeling clay to form the missing piece. Metric dimensions could also be requested. Then compare their product with the real piece.

One final space-time challenge—find the smallest possible box in which the log puzzle pieces can be stored.

Additional Considerations on Space and Time

Think of the many opportunities for using the skill of distinguishing spatial and temporal relationships. This skill is essential for the young child's early learning—such as the initial recognition of familiar shapes and places, the gradual development of a sense of temporal causality, the awkward beginnings and steady development of all sorts of motor coordination. During our daily lives, spatial and temporal relationships must be distinguished in driving a car (and parking it), cleaning a room, painting a house, mowing the grass, shoveling snow, preparing a meal. A great many of your decisions in the classroom will also center on the use of spatial and temporal relationships.

This skill is certainly not limited to astronauts' determining a landing site, or to archeologists' identification of prehistoric bones situated in certain positions at specific depths in the earth, or to physicists' calculations of the half-lives of radioactive elements being bombarded within nuclear reactors. Each of us is constantly presented with countless opportunities to make better decisions based on the ability to distinguish spatial and temporal relationships. This is a skill well worth constantly nurturing and practicing.

Classifying

How many times has an intended few seconds' search for something ended up becoming an exasperating, near-endless exploration? Why is it so easy to locate items in some stores and so frustrating in others? What's the crucial ingredient in these situations? It's the skill of classifying, or the ability to make order out of disorder.

Scientists have developed numerous classification systems—plants and animals; solids, liquids, and gases; elements, compounds, and mixtures; acids and bases; sedimentary, igneous, and metamorphic rock, to mention just a few. Such systems are based more on arbitrary decisions than on natural laws. They are designed to facilitate communication or to organize information so that it will be more understandable and more retrievable. For instance, suppose you heard that a new species of amphibians was recently discovered along the Amazon River. If you have had previous experiences or knowledge of other amphibians (like frogs, toads, salamanders, or newts), you would already have a fairly accurate idea of what this newest member of the group is probably like. Classifying is also important in concept formation.

Activities

Selecting Similar Sets

A. Introduction   There are innumerable ways of grouping objects, but the most obvious criterion is the objects' physical properties. In this activity you can investigate various means of establishing order out of disorder by grouping objects into sets with similar characteristics.

B. Materials   (1) A hodgepodge collection of fifteen to twenty varied pencils (select the almost-forsaken ones that seldom are used and would probably never be missed) and (2) a metric ruler.

C. Procedures

1. How many physical properties can you identify among the assortment of pencils? What other characteristics of pencils can you add to the initial list? Examples:

   Plain or figured
   With or without erasers
   Smooth or nicked
   Long or short
   Round or flat-sided
   With or without points

2. Suppose you've just been hired as a pencil pusher. The position has the following job requirements:
   a. Bring only yellow pencils on Monday.
   b. Tuesday's pencils must be over 12 cm long.
   c. Wednesday's pencils must have 5 mm of lead exposed.
   d. Only pencils with 15 or more letters imprinted are allowed on Thursday.
   e. Friday is clean-up day; all edges of pencil erasers must be at least 5 mm long.
3. Prove that you can master these job requirements. Group your pencils to meet each day's requirement. That is, first separate the pencils into two groups: a set of yellow pencils and a set of not-yellow pencils. Then form two new sets, one consisting of pencils longer than 12 cm and the other of pencils shorter than 12 cm—and so on.
4. Practice grouping the pencils into similar sets by using other physical properties. How many different combinations of similar sets can you determine? Share the results of your ideas with another pencil-pusher.

D. Extensions   There are countless variations on this theme of classifying according to physical characteristics. Consider using any of the following:

For buttons

Two holes versus not–two holes
Plastic versus not-plastic
White versus not-white
Smaller than a dime versus larger than a dime

For seeds

Light-colored versus dark-colored
Hard versus soft
Rough surface versus smooth surface
Able versus unable to "float" in air

For leaves

Jagged edges versus not-jagged edges
Shorter than 10 cm versus longer than 10 cm
Rough versus smooth
With insect holes versus without insect holes

For people

Male versus female
Brown eyes versus not-brown eyes
Long hair versus short hair
With glasses versus without glasses

Objects could also be classified into other categories of physical properties such as sinkers or floaters, magnetic or not-magnetic, symmetrical or asymmetrical, flexible or rigid.

Another common type of criterion for grouping objects is use. Think of the general categories apparent in most department stores or large discount stores. Items can be classified in terms of possible uses: for play, for building, for cooking, for making music, for sending messages, for keeping healthy, and so on.

When working with children, bear in mind that the younger the children, the greater their need to physically handle the objects and sort them into visible, tangible groups. Encourage very young children to put objects of similar sets within circles of yarn or into plastic buckets or any other suitable container. One final suggestion: Whenever one quality is compared with another, a standard for comparison is needed. For example, when is a seed considered to have a light color or a dark color? What determines roughness and smoothness? Encourage the children to decide upon their own standards before they begin to group items, but be prepared for a few arguments.

Sorting Systems

A. Introduction In today's technological society, people are constantly looking for ways to make work easier. Many classifying tasks can be carried out by machines. There are machines to sort coins or eggs according to size, to separate liquids according to content and density, to sift gravel or particles of minerals, and to separate magnetic materials from nonmagnetic junk. Perhaps you can think of additional examples of shortcuts for grouping objects or materials according to their physical properties.

A computer is one machine that can sort a tremendous amount of information in a brief period of time. In this activity you can develop your own modified computer sorting system. It can become a time-saving device and provide a wealth of information with the flick of a paper clip.

B. Materials (1) At least ten to twenty small index cards; (2) a hole puncher; (3) scissors; and (4) a paper clip three-fourths straightened out.

C. Procedures:

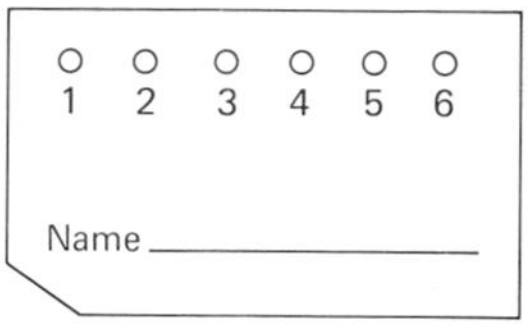

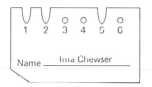

Figure 5-4 Sorting card.

1. First prepare the index cards as shown in Figure 5-4. Punch as many holes as you wish; then number them consecutively. *Be consistent:* every card should be identical initially. Space the holes evenly, about 5 mm from the top edge of the card. Cut off a small piece from the lower left corner. (Regular computer cards could also be used.)

2. Next, decide what kind of information you want to collect, sort, and analyze. Again, there are endless possibilities, but perhaps some examples of a brief preference survey might help. You might try asking questions like the following to a variety of persons:
   (1) Do you prefer to be indoors or outdoors?
   (2) Would you rather be alone or with people?
   (3) Do you have a four-legged pet?
   (4) Should space exploration be increased?
   (5) Should science be taught to children in the primary grades?
   (6) Should the school year be 12 months long with 1 month off every 3 months?

3. If a person answers, "Yes," or selects the first choice, the paper above the corresponding punched hole should be cut away to make a wedge-shaped opening. If a negative response or the second choice is selected, no wedge is cut out. Figure 5-5 shows that Ima Chewser (1) prefers being indoors, (2) prefers being alone, (3) does not have a four-legged pet, (4) would not increase space explorations, (5) does think science should be taught in the primary grades, and (6) does not want a 12-month school year.

Figure 5-5 Ima's choices.

4. After responses have been collected from a variety of people, arrange the cards in a stack so that the numbers and holes are superimposed. (The lower left corner should appear clipped from the entire stack.) Then run the paper-clip wire through the first hole. All the cards that fall from the wire indicate a positive answer or selection of the first choice. The cards remaining on the wire indicate a negative answer or selection of the second choice. Continue these procedures with each question and the corresponding punched hole. This sorting system gives an almost instant classification of responses.

5. Now design your own set of questions to be sorted. Use as many questions as you wish or as can be fitted around the edges of the card. The questions

must require either a "yes" or a "no" answer or one of two choices. Then select the most appropriate number and kinds of respondents. Once all the wedges have been cut out, sort the results by means of the paper-clip technique. Then analyze the results.

D. Extensions   This punch-card–paper-clip technique can be modified for a wide variety of uses.

For example, it can be a means of keeping track of the number of lessons or activities students have completed. That is, students could be instructed to cut out a wedge above a hole after completing a specific lesson, activity, or evaluative task. Within several flicks of the paper clip you could sort your class's progress.

Cards could also be made with predetermined cuts and used to identify various kinds of plants, animals, or minerals, or a wide variety of information. For example, examine the three leaves shown in Figure 5-6 and then use the sample cards to identify each leaf. You could also include many more descriptors for any number of specimens.

Encourage students to design their own sorting machines, or techniques for separating mixtures. For example, how might a mixture of sawdust, fine sand, and salt be separated? Suppose someone knocked over a bag of jelly beans into a rice container, which then fell into a sugar canister. How can this mess be separated?

Figure 5-6   Using sorting cards to sort leaves.

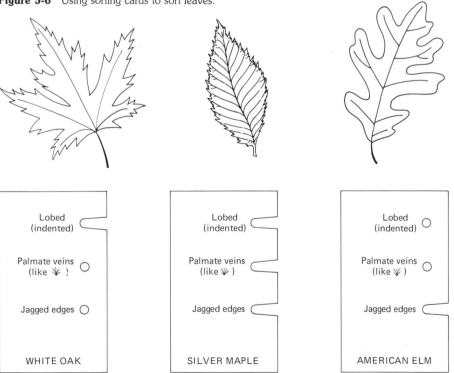

Leafing through Stages

A. Introduction There are no set laws or codes for classifying. There are, however, some basic considerations and techniques. In order to group objects in any kind of orderly fashion, you first need to consider their attributes and values. An *attribute* is a distinct yet general characteristic, such as shape, size, color, texture, composition. *Values* are specific variations of attributes. For example, the values for the attribute *shape* could be *circle*, *ellipse*, *triangle*, *square*, *rectangle*, *pentagon*, and so forth. Values for the attribute *color* include *red*, *green*, *yellow*, *brown*, and *chartreuse*. Two basic values of *texture* are *rough* and *smooth*, with countless variations in between. Values for the attribute *composition* include *paper*, *wood*, *plastic*, *metal*, *glass*, and *leather*.

One of the most common uses of attributes and values can be found in supermarkets. The aisle signs—such as CANNED VEGETABLES, PAPER PRODUCTS, CLEANING MATERIALS, SNACKS—are attributes designating where a variety of their values can be found. Examine Figure 5-7: suppose that a

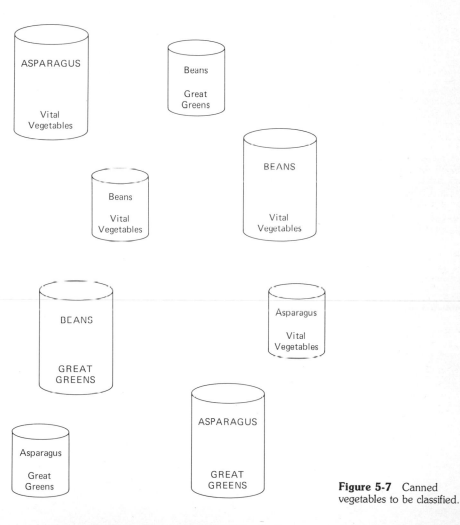

Figure 5-7 Canned vegetables to be classified.

small sample shipment of canned vegetables has just arrived in a supermarket, as shown. The cans are to be arranged according to some sort of classification scheme. What attributes and what related values should be considered? Examine the following list:

Attributes

Kind of vegetable
Size of can
Brand

Values

Asparagus or beans
Large or small
Vital Vegetables or Great Greens

One possible scheme for classifying these attributes and values is shown diagramatically in Figure 5-8. This arrangement represents a three-stage classification scheme. The first stage consists of the attribute *kind of vegetables* with two values, *asparagus* and *beans*. Of the total of eight canned vegetables, four are asparagus and four are beans. The second stage, a subdivision of the first stage, shows the attribute *can size* with the values *large* and *small*. There are two large cans of asparagus and two small cans. There are also two large cans of beans and two small cans. The third stage, a further subdivision of the second stage, consists of the attribute *brand* with the two values *Vital Vegetables* and *Great Greens*. Note that all eight cans are accounted for in each stage. (If an object did not fit with any of the values identified, another value would have to be selected.) In this activity you will be able to practice identifying different attributes and values for organizing a four-stage classification scheme.

Figure 5-8 Three-stage classification scheme.

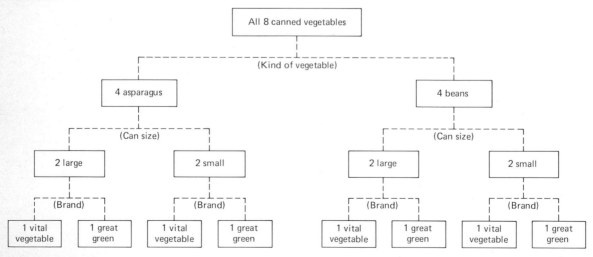

B. Materials   Only Figure 5-9.

C. Procedures
1.   Compare the leaves shown in Figure 5-9. What general characteristics or attributes do all the leaves have? The most obvious attributes are size, edges, veins, and holes. (There are other attributes, like the number of veins

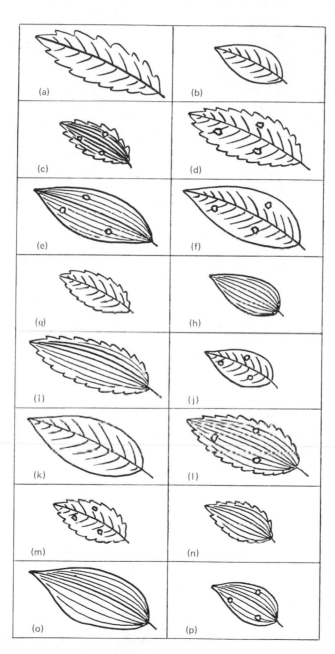

Figure 5-9 Leaves for four-stage classification scheme.

Table 5-3

ATTRIBUTES	VALUES
1.	
2.	
3.	
4.	

shown, the length of the petiole or stem, and the number of indentations along the edge. But size, edges, veins, and holes are more apparent and all-inclusive.)

2. Now determine the values that correspond to each attribute. It might be helpful to record your observations in tabular form, as shown in Table 5-3.

3. Fill in the attributes and their corresponding values in Figure 5-10. Also, include the letters of the leaves in the appropriate boxes.

4. When you finish, you should be able to name the four values for each leaf in the fourth and final stage of the classification scheme.

5. The classifications of the canned vegetables and the sixteen leaves are contrived examples. Exceedingly few classifications of objects from the natural world will ever work out so symmetrically. Select a new collection of materials, or use any of the items previously suggested—such as pencils, buttons, real leaves, or even the people around you. Design your own original single-stage classification scheme, then a two-stage scheme, a three-stage scheme, and a four-stage scheme. You might even try going beyond the four-stage scheme.

Figure 5-10 Four-stage classification scheme.

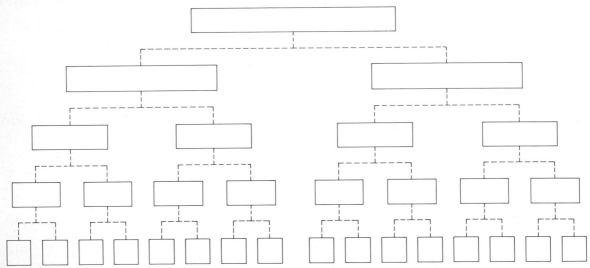

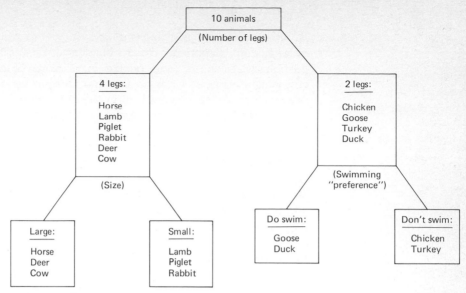

Figure 5-11 Example of an asymmetrical classification scheme.

D. Extensions   There is no rule that requires the use of the *same* attributes and values for both sets after the initial grouping. Suppose one were classifying the following ten animals: horse, chicken, lamb, goose, piglet, rabbit, turkey, deer, cow, and duck. The first stage might be based on the attribute *number of legs* with the values *four legs* and *two legs*. The four-legged animals might next be grouped according to the attribute *size* (*large* or *small*), whereas the two-legged animals could be grouped according to the attribute *swimming preference* (*does swim* or *doesn't swim*). Figure 5-11 illustrates this sort of unbalanced or asymmetrical classification scheme.

The first stage need not always consist of just two groups. Suppose people in a room were to be classified according to the attribute *hair color*. Instead of using the values *brunette* or *not-brunette*, six values could be used: *brunette, black, blonde, red, gray,* and *multicolored.* A single-stage classification could begin with any number of values. However, the more values you select initially, the more characteristics you will have to deal with in each succeeding stage.

When presenting the skill of classifying to children in the primary grades, begin with single-stage schemes which deal with just two values. Gradually, the children may be able to carry out two-stage classification schemes with your help. Remember that it is extremely difficult for preoperational children to attend to many ideas simultaneously, and it is virtually impossible for them to reason logically or systematically. Older or more advanced children in the concrete-operational stage usually can perform multistage classification schemes, but only with real objects. Diagrams such as Figures 5-8, 5-10, and 5-11 can facilitate the recording of information or keeping track of the members of each set.

Pencil-Pen Pickup

A. Introduction   Classifying is an extremely important component of the skill of analyzing. Reporters, clerks, builders, police officers, lawyers, doctors, economists, scientists (and people in all sorts of other occupations) must constantly classify all kinds of data and ideas in various terms: fact or fiction, relevance or irrelevance, and so on. All of us probably do far more classifying than we realize. Sometimes we may go about it systematically; at other times, haphazardly. In this activity you can develop some deliberate strategies for classifying in order to pick up the correct pencil or pen. (You'll need at least one partner to complete this activity.)

B. Materials   (1) A hodgepodge collection of ten to fourteen ballpoint pens, and (2) the collection of pencils that were used in "Selecting Similar Sets" (see page 123).

C. Procedures
1. Examine the assortment of pencils and pens. If any two are *exactly* identical, eliminate one of them. One person should *mentally* select a specific pen or pencil.
2. The other person or persons should then ask questions of the person who chose the pencil or pen which can be answered only by "yes" or "no."
3. Keep a record of the number of questions needed to identify the pencil or pen.
4. Repeat these procedures but this time devise a plan to ask as few questions as possible to identify the object more quickly and efficiently. A helpful hint for your plan can be derived from reexamining the general pattern of the classifying scheme illustrated in Figure 5-10. Either physically or mentally separate the nonchoices from the remaining possibilities. Deliberately try to eliminate as many possibilities as you can with each question.
5. By using your classifying strategy, how many questions did you ask before demonstrating the correct pencil or pen pickup?
6. Reverse roles and try out different strategies. Which classifying strategy was the most efficient?

D. Extensions   Substitute any other items for the pencils and pens. For children, use objects which can be easily moved about into groups, since children in the preoperational or concrete-operational stage usually will not be able to construct the classification schemes mentally. The activity could also be simplified by using fewer objects.

Another game-like activity involving classifying is "Guess My Rule." Use any random set of materials that have rather obvious attributes. Diagram a multistage classification scheme showing the values used but not the attributes. Ask someone to identify the appropriate attributes. Or reverse the procedures—specify the attributes on the diagram and ask another person to select the values.

Additional Considerations on Classification

You've experienced the skill of classifying as a means of establishing order by grouping objects according to similar and dissimilar properties. I hope you've also discovered many new arrays of interrelationships. But the most fundamental use of classifying is often overlooked—its role in concept formation.

A concept is a general idea developed by grouping together a set of common characteristics. Concepts evolve from classifying common stimuli or experiences. A young child, for instance, does not immediately attain the concept "dog" or "dogness." For quite some time, a teddy bear, kitten, duckling, or even a heavily bearded, curly-haired man might be considered a "doggy" by a young child. Only as the child sorts out or *classifies* the key attributes of dogness does the concept become more generalizable and, therefore, more valid. Review Piaget's interaction model again, especially the process involving the initial sorting through recall. The process of classifying is an integral part of *all* learning—continuing from infancy until the last thought.

A concrete example might be useful. Examine Figure 5-12: one of the best

All of these are *Norleys*.

None of these is a *Norley*.

Which of these are *Norleys*?

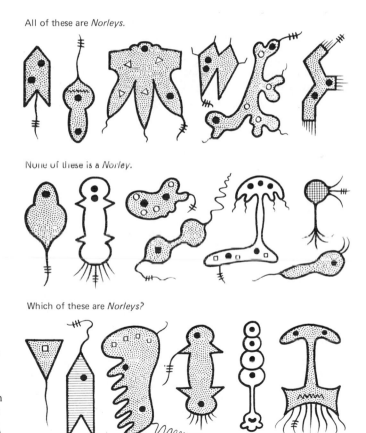

Figure 5-12 Fictitious concept formation. (Copyright © 1966, 1968, by Educational Development Center. All rights reserved. Reproduced by permission of the Elementary Science Study of Education Development Center, Inc.)

assortments of examples comes from an Elementary Science Study unit, *Attribute Games and Problems* (1962, pp. 9–47, 63–77). A portion of this unit deals with the identification of fictional concepts by means of "creature cards." Figure 5-12 illustrates one of the fifteen creature cards, the "Norley."

To really know a Norley when you see one, you must be able to sort out its key characteristics or attributes from unrelated traits. A Norley has three identifying attributes: tiny dot shading inside, two black larger dots, and at least one tail with three short parallel perpendicular lines. Each of the nonexamples lacks at least one of these attributes; some lack two. None has the combination of all three. One of the most efficient ways of determining if children have comprehended a concept is to give them the opportunity to select (or sort) examples of the concept from nonexamples.

This technique need not be used with just fictitious creatures. *Any* concept can be presented in this way if you have the ability to identify its key attributes. An example dealing with a real-life concept is presented in Figure 5-13.

Figure 5-13   Concept identification.

All of these are scallops.

None of these is a scallop.

Which of these are scallops?

One of the dangerous results of inadequate concept formation (or of limited or incomplete classification of experiences) is the formulation of stereotypes. Stereotypes are overgeneralizations or rigid concepts formed by people who are impervious to certain kinds of experiences. They result from inaccurate or incomplete assimilations of, and accommodations to, encounters. Before moving on to the next process skill (communicating), do some sorting on your own tendencies in concept formation, and examine how you use the skill of classifying in your own living and learning.

Communicating

Oral transmission of information is, obviously, the most common and immediate form of communication used by most people in their everyday living. Perhaps not so obviously, it is also the most common form of communication used by scientists.

Activities

"Is It This One?"

A. Introduction If a listener can identify or select an object you describe, you'll know that you've communicated well. The more accurate your description, the less opportunity for confusion.

This is a group activity for three to five participants. You should be able to describe an object so clearly and accurately that another person can identify it on the first try.

B. Materials (1) A set of leaves from six different common trees in your area—one leaf from each different tree. One possible set of leaves is shown in Figure 5-14 (page 136). In case you're interested, the names of the leaves illustrated are (a) sweet gum, (b) American elm, (c) pin oak, (d) hackberry, (e) Chinese elm, and (f) silver maple. Select your set of leaves according to the abilities of your students—that is, use less similar leaves for younger students and more-similar leaves for older or more capable students. (2) Also, collect a set of six or more leaves from the same tree.

C. Procedures
1. Imagine that you've just received word that you can win a tree. All you have to do is phone in a description of your selection. When the tree planters arrive with your choice, they will ask, "Is it this one?" If you can respond, "Yes," they will plant the tree for you.
2. Examine the collection of leaves carefully.
3. Ask the other participants to either close their eyes or look elsewhere as you describe (without naming) one of the leaves. (If they watch you while you're

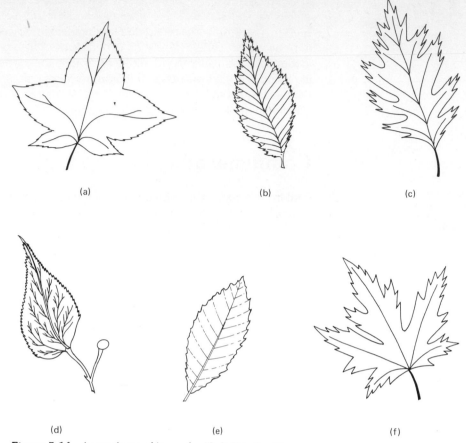

Figure 5-14 A sample set of leaves for "Is It This One?"

describing, your eyes might give away too many clues. Also, careful listening is a vital component of good communication.)

4. After you have completed your description, the participants should be able to identify your choice and ask, "Is it this one?" (If there was poor communication, determine the reasons for it.)

5. Exchange roles until at least four of the leaves have been accurately described and selected.

6. Next, examine the set of six or more leaves from the same tree and repeat procedures 3, 4, and 5.

7. Try two variations: (a) Make your description as long as you can and determine if you have still communicated; then (b) make your description as brief as possible and see if the correct leaf is selected. Discuss the results of the two methods.

D. Extensions   Try substituting a variety of materials such as seeds, shells, rocks, potatoes, pencils, or rubber bands.

For children, you might ask each child to place one of his or her shoes in a group and then have a child describe one shoe—the other children will select one and ask, "Is it this one?" Then, make the game harder by asking each child to place both shoes in the pile.

Another game-like activity, "Information, Please," also improves oral communication. A specified assortment of objects, or any visible objects in a specified area, can be used. Only attributes or characteristics should be given—never the actual name of the object. One person begins with a statement such as, "I am thinking of a rectangle." If the others believe more information is needed to identify the object, they say, "Information, please." The purposes are to encourage accurate, thorough descriptions and to discourage wild guessing.

Discuss examples of poor and good oral communication. What are some of the difficulties involved in oral communication? When is it extremely important that the oral communication be clearly understood? How might oral communication best be improved?

Do-It-Yourself Picture Power

A. Introduction   Diagrams, illustrations, and photographs are often used to communicate procedures and information. Children sometimes find drawing an easier means of communicating than speaking and far more enjoyable than writing. Even many adults cannot grasp a problem until they not only mentally but also physically picture the parts. Clear pictorial representations can also communicate a wealth of information almost instantly. However, if you have ever attempted to assemble children's toy kits by following the illustrated directions, you will realize that pictures don't always communicate clearly. In this activity you'll have an opportunity to determine what kind of power you can derive from certain pictures.

B. Materials   All the items shown in Figure 5-15 (page 138).

C. Procedures

1   Assemble the parts shown in Figure 5-15.
2.   Then release the assembled system, preferably on a smooth surface.
3.   Evaluate the results. How clearly did the pictures communicate? How well did your results conform to their messages? What additional pictures or information might be needed?

D. Extensions   The small wooden rod sometimes has a tendency to slip. Either tape it securely or hammer a *small* nail next to it to serve as a barrier. Try using different rubber bands to create more "go power."

The "crawling spool" was a common homemade toy in colonial days. Survey your relatives, neighbors, or older acquaintances to determine their

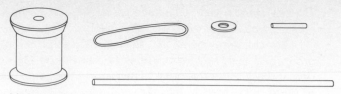

(a) Components for do-it-yourself picture power

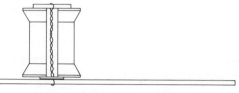

(b) Inside view of the assembled parts

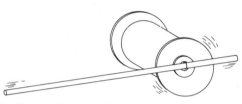

(c) The assembled parts "on the go"

Figure 5-15 "Do-it-yourself picture power."

familiarity with this type of toy. Inquire about possible modifications or other homemade toys they might have used. Try to use their ideas.

If possible, analyze the illustrated directions for assembling commercial toys or various types of equipment. Evaluate the adequacy of the illustrations. What improvements in communication would you recommend?

Design your own toy, tool, or technique. Sketch an illustration of it. Ask someone to interpret your drawing and duplicate your product. How well were you able to communicate pictorially? If you need some ideas, investigate the Nuffield Science 5-13 series, especially the unit *Science from Toys*—Stages 1 and 2 and "Background" (1973). This entire series is also an excellent communication of a Piagetian-based science program.

Following Directions

A. Introduction   It may seem superfluous to discuss the importance of the printed word as a means of communicating. However, this form of the skill always needs to be reinforced at all levels of learning—especially as regards following directions.

B. Materials   Figure 5-16 and a metric ruler.

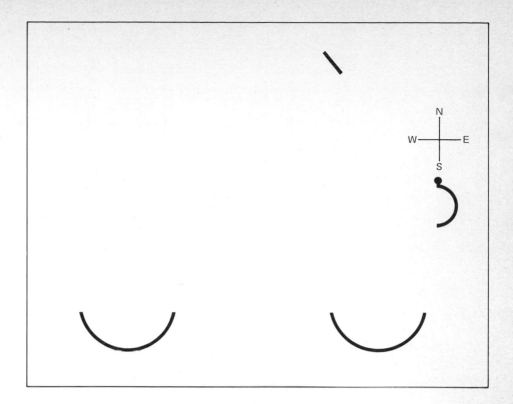

Figure 5-16
Following directions.

C. Procedures Begin at the dot below the S in Figure 5 16. Draw line segments for the following lengths:

1. 3 cm west
2. 3.5 cm north
3. 5 cm west
4. 3 cm south
5. 1.5 cm west
6. 3 cm south
7. 5 mm west
8. 1 cm south
9. 10.5 cm east
10. 1 cm north
11. 5 mm west
12. 3 cm north

If your final result reminds you to move along, communication has occurred.

D. Extensions If you use the preceding activity or any modification of it with children, it might be helpful to use centimeter grid paper.

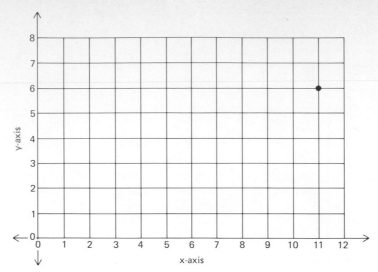

Figure 5-17 Coordinates.

A *coordinate system*—that is, the use of numbered pairs of lines to locate precise points of intersection—is excellent for communicating activities that have to do with following directions. If you are slightly stale on coordinate techniques, some refreshing might be in order. There are some important rules that must be followed in order to maintain consistency and clear communication. Study Figure 5-17, and note that the horizontal line is called the *x-axis* and the vertical line, the *y-axis*. In order to identify the position of a point of intersection of the axes—that is, the *coordinates* of that point—always state the x-axis reading first and then the y-axis reading. Place a comma between the two numerals and enclose both in parentheses. For example, the intersection point or coordinates illustrated in Figure 5-17 is (11,6).

Here's another sample you can try, which employs a coordinate system as shown in Figure 5-17. Locate the points for each pair of the following coordinates. Then draw a connecting line segment between the two points. Connect the following pairs of coordinates:

1. (7,4) and (1,4)
2. (1,4) and (1,0)
3. (8,7) and (9,7)
4. (8,2) and (9,0)
5. (11,5) and (11,6)
6. (1,0) and (3,0)
7. (7,6) and (7,3)
8. (2,2) and (3,0)
9. (11,5) and (9,5)
10. (8,2) and (8,3)
11. (9,0) and (7,0)
12. (7,3) and (8,4)

13. (7,0) and (7,2)
14. (7,6) and (8,7)
15. (9,5) and (8,3)
16. (2,2) and (7,2)
17. (8,6) and (8,4)
18. (9,7) and (10,6)
19. (10,6) and (11,6)
20. And don't forget (1,4) and (0,6)

You may not be panting with pride, but your product should resemble something that could. How could the directions be communicated more clearly? Try designing a clearer set of directions for constructing a different figure.

It is not always easy to communicate by means of the printed word. Because of the need for clarity and specificity, both the writing and the reading of highly specific reports can be quite tedious. To increase your empathy with those who must use this sort of communication, try writing directions for a hypothetical Martian on how to tie a shoe, drive a car, prepare a meal, or use a microscope.

Getting a Grasp on Graphing

A. Introduction   Graphing always seems to be one of the most "scientific" modes of communicating. For anyone who feels "I'm not the scientist type," it's also one of the least understood. Forbidding as it may appear at first, a graph is only a pictorial representation of related data. It's simply a means of showing the relationship between two sets of information or two factors.

Graphing can be introduced quite clearly and simply, as, for example, in Module 25 of *Science—A Process Approach II* (1974). Children are asked to physically stack cubes according to their colors. For instance, they might construct a vertical column of two green cubes, five red cubes, and so on. They next establish one-to-one correspondence by placing similarly colored felt squares on a felt board. Either the bottom edge of the felt board or a string aligned horizontally can serve as the baseline. A graph of the felt-board arrangement is then drawn on the chalkboard or on a large sheet of paper, or shown from an overhead projector. Finally, the children are asked to draw their own picture or graph of the results, as shown in Figure 5-18. In this series of three to five activities, the children progress from concrete to pictorial to a more symbolic type of understanding of graphs.

Figure 5-18 shows a bar graph, or the representation of *discontinuous factors*. In other words, the colors are separate, discrete considerations; there are no in-between qualities among them. The graph simply shows the relationship among the numbers of different-colored cubes—two green, five red, three yellow, and six blue cubes, and one orange cube. Other examples of bar graphs showing discontinuous factors could include the number of kinds of cars you saw today or the number of people in a group having red, blond, brunette, black, or gray hair.

Graphs can also show the relationships between *continuous factors* such as growth over time (see Figure 5-19), or distance moved over time, or number of

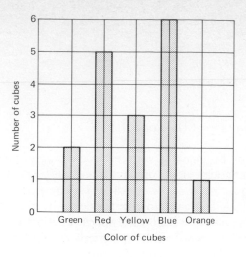

Figure 5-18 Sample bar graph (discontinuous factors).

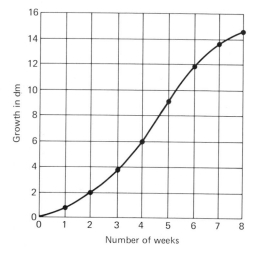

Figure 5-19 Growth of a plant (continuous factors).

masses added and amount of stretch of a rubber band or spring. In Figure 5-19, a continuous relationship is shown between the first week, the second week, the third week, and so on. A line graph best illustrates a continuous relationship.

Just as there are certain conventions to be used with coordinates, so too are there specific rules for constructing graphs:

1. The numerals along the vertical line (or along the horizontal line when appropriate) should be of equal increments and should match up with the lines of the grid rather than with the spaces between the lines.
2. Both axes, the vertical and horizontal, always should be labeled with the appropriate units or descriptors.
3. The data representing the results of the events or the measurements recorded (that is, the responding factors), should be placed on the vertical axis.

4. The data which are automatically changing or systematically being changed (for example, the number of seconds, hours, or days, or the number of equal masses being attached to a spring) are placed on the horizontal axis.

In other words, the horizontal axis shows what was manipulated; the vertical axis shows what results were observed. Experience will enhance your understanding of these principles. By completing this activity, you should get a better grasp on graphing.

B. Materials (1) A bag of twelve or more golf balls (or any similar type of balls); (2) a 12-in. or 30-cm ruler which has an indented center groove down the entire length; (3) about 1 dm of masking tape; (4) a meter stick or metric tape; and (5) a marble.

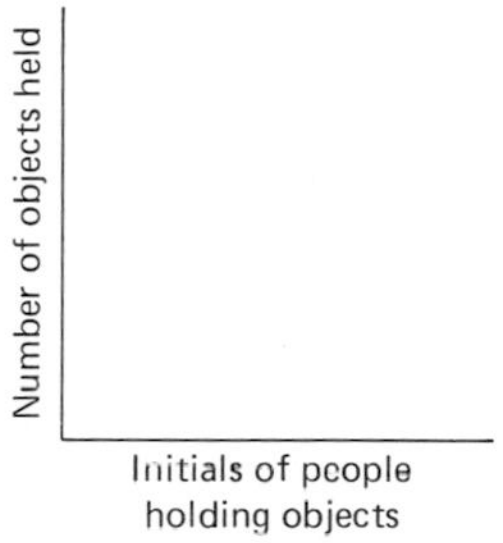

C. Procedures

For bar-graphing discontinuous factors

1. Select five or more people of varying sizes or ages to help you. Make a list of their initials.
2. Separately, ask each person to reach into the bag and lift out a handful of golf balls. Record the results next to their initials.
3. Graph the data as suggested in Figure 5-20.
4. Review the preceding rules on graphing and the descriptions of discontinuous factors. In what ways are you getting a better grasp on graphing?

For line-graphing continuous factors

1. Find a fairly open area which has smooth carpeting. (If no such area is available, use a rather long table covered with a tablecloth or bed sheet.)
2. Attach about 5 cm of the total 10-cm length of masking tape to the *back* side of the ruler at the 1-cm end. Fold back the remainder of the unattached masking tape to the edge of the ruler, sticky side exposed. Then firmly press the tape onto the working surface previously described. The ruler should now be like a hinge which can be moved only up and down.
3. Place a book or any sturdy object under the unattached part of the ruler so that the ruler is lifted about 8 cm. At this point you should have a ramp which has about a 15° angle.
4. Line up the meter stick or metric tape near and even with the attached end of the ruler so that it is parallel to the direction a marble would roll from the ramp. Compare your setup with Figure 5-21.
5. Practice *gently* placing the marble in the groove and releasing it from various points on the ruler. So that the marble can roll freely from the ramp, try to eliminate any irregularities—such as bumps in the path, too-close

Number of objects held

Initials of people holding objects

Figure 5-20 Grasping graph.

Figure 5-21 Ruler-ramp marble roller.

8 cm

proximity of the meter stick or tape which hinders the movement of the marble, and inconsistent releases of the marble.

6. Measure the distance the marble rolls when released from the following points on the ruler ramp: 5 cm, 10 cm, 15 cm, 20 cm, and 25 cm. Since it is practically impossible to consistently control the release of the marble, take at least three to five measurements from each release point and determine the average roll distances. Keep a record of the results of all your measurements.

7. Plan how to graph your data: the release points on the ruler ramp and the distances rolled from the end of the ramp. Which data should go on the horizontal axis and which on the vertical axis? If necessary, reexamine the graphing rules on pages 142 and 143.

D. Extensions   Compare the graph you constructed with the graph shown in Figure 5-22. In what ways are they similar? How do they differ? How do both graphs show the relationships between continuous factors?

The actual constructing of a graph is not difficult once you understand how to identify the axes and their corresponding data. But you also need to be able to interpret graphs. The data from bar graphs, or discontinuous factors, are to some extent self-evident. But it might be helpful to look more closely at line graphs

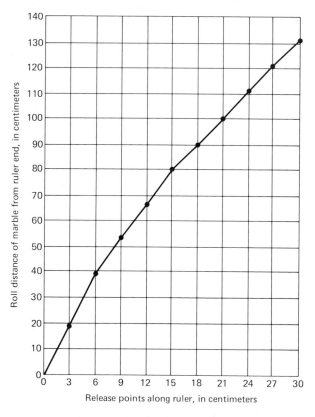

Figure 5-22   Distances rolled by marbles from a ruler-ramp.

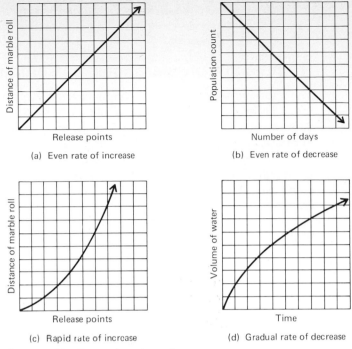

Figure 5-23 Types of graph trends.

showing continuous factors. Study the hypothetical graphs of Figure 5-23. These represent some of many possible trends:

Graph A shows a continuous, even rate of increase—for instance, if your marble rolled 10 cm farther after being released from each successively higher point along the ruler-ramp, that would be an even rate of increase.

Graph B shows a continuous even rate of decrease. Suppose you have a sample of pond water containing 1000 tiny creatures on day 1. Each day you sample the water to ascertain the population count. On day 2, you find only 900 creatures; on day 3, you find 800, on day 4, you find 700; and so on. This illustrates a constant decrease in population.

Graph C shows a rapid rate of increase comparable to a doubling effect. Suppose your marble rolled 10 cm when released from the 3-cm point, then 20 cm when released from the 6-cm point, then 40 cm from the 9-cm point, 80 cm from the 12-cm point, 160 cm from the 15-cm point, 320 cm from the 18-cm point, and so on. The distance rolled is increasing rapidly.

Graph D exhibits a gradual rate of decrease. For instance, if water were leaking from a tiny hole in the bottom of a car radiator, the amount of leakage would gradually decrease over time.

Now try to interpret the graph you constructed for your data from the roller-ramp. Which graph from Figure 5-23 does your graph most closely match? Think of some additional examples that correspond to the other graphs.

Additional Considerations on Communication

There are numerous modes of communication available to us—spoken words, printed words, three-dimensional symbols and models, pictures and graphics of all sorts, electronic impulses, hand signals, and symbols conveyed by touch, as in Braille. Although the means may vary greatly, the goal is always the same—to get a message across, or—even more fundamentally—simply to share. Scientists constantly strive for clearer, more accurate communication. Society as a whole could profit by doing the same. Think of the many times you use the skill of communicating during the course of a normal day. Or think of the times you needed to communicate but for one reason or another felt unable to do so. Children, too, need to communicate. They also need the freedom to use their own idiosyncratic means of communicating at times. Before moving on to the next chapter, think about ways of incorporating and improving the skill of communicating in your own life and in the lives of children. Also, think how communicating is integrally related to the framework of observing, measuring, distinguishing spatial and temporal relationships, and classifying.

Encounters

1. What do you consider to be the most important reasons for emphasizing the process skills in elementary science?
2. Try with children any of the four main activities described for each of the three process skills (distinguishing spatial and temporal relationships, classifying, and communicating). What modifications were necessary? How might the activities be improved?
3. Develop activities for any of the ideas suggested in the "Extensions." (You have over thirty choices.) Share your products and results with your colleagues.
4. Suppose you are presenting an activity to children on distinguishing spatial and temporal relationships, or classifying, or communicating. The principal walks into the room and asks, "What's the point of this lesson?" You respond, "The children are learning to distinguish spatial and temporal relationships (or to classify, or to communicate)." The principal then asks, "What does that mean?" What responses would you give to describe each of these process skills?
5. Make a list of the instances in your daily life when you distinguish spatial and temporal relationships, classify, and communicate. How might you improve the use of these skills?
6. Examine some elementary science textbooks and teachers' guides. Find examples of the use of the three skills discussed in this chapter. Also find, instances where these skills were *not* used. Then develop activities related to these situations for incorporating the process skills discussed in this chapter.
7. Your choice.

References

Attribute Games and Problems, Elementary Science Study, Webster Division, McGraw-Hill, New York, 1974.

Module 25: "Introduction to Graphing" (Communicating/b), in *Science—A Process Approach II,* American Association for the Advancement of Science, Ginn and Company, Lexington, Mass., 1974.

Science from Toys—Stages 1 and 2 and Background, Science 5/13, Schools Council, Nuffield Foundation and Scottish Education Department, Macdonald-Raintree, Milwaukee, Wisc., 1973.

Suggested Readings

Butts, David P., and Gene E. Hall: *Children and Science: The Process of Teaching and Learning,* Prentice-Hall, Englewood Cliffs, N. J., 1975. (Chapters 3, 5, and 10 give additional ideas for activities related to the process skills presented in this chapter.)

Commentary for Teachers: Science—A Process Approach, American Association for the Advancement of Science, Ginn, Lexington, Mass., 1970. (This is an excellent resource book on the thirteen process skills used in the elementary program *Science—A Process Approach.*)

DeCecco, John P.: *The Psychology of Learning and Instruction: Educational Psychology,* Prentice-Hall, Englewood Cliffs, N. J., 1968. (This text presents a very lucid treatment of defining concepts and recognizing the nature of concepts and their relationship to principles, and steps for teaching concepts. This information will help clarify the importance of classifying.)

DeVito, Alfred, and Gerald H. Krockover: *Creative Science: Ideas and Activities for Teaching and Children*, Little, Brown, Boston, 1976. (Over 100 activities are presented for utilizing the process skills and a potpourri of science concepts. There is a special section entitled "Shoestring Sciencing" which provides numerous ideas for using inexpensive materials.)

Preservice Science Education of Elementary School Teachers, American Association for the Advancement of Science Commission on Science Education, Washington, D.C., 1970. (The third guideline of this important document stresses the need for competence in the processes of scientific inquiry. Twelve specific skills are described.)

Theory into Action . . . In Science Curriculum Development, National Science Teachers Association (NSTA), Washington, D.C., 1964. (Through this document, NSTA, the largest professional organization for science teachers, offers some useful guidelines for evaluating the directions of science education. The second section deals with identifying some of the major conceptual schemes and processes of science.)

CHAPTER 6
GOING BEYOND
THE BASICS

Much of the information derived from observing, comparing, measuring, distinguishing spatial and temporal relationships, classifying, and communicating fits well under the general notion of empiricism. For those who believe that science is no more or no less than the collective body of facts derived from sensory experiences, these actions are logical and, taken together, sufficient. However, what we often fail to realize is that a vast proportion of the scientific endeavor involves going beyond basic skills. Much of science is based on working from one's best hunches, or attempting to interpret a multitude of observations so as to discover relationships.

No one has ever traveled to the center of the earth or beyond the solar system, yet there are volumes of descriptions on what such journeys might be like. The same is true of how our planet, or our solar system in general, was formed, and of how life began and evolved; though no one saw these processes, many people have described them. One of the most astonishing discoveries a nonscientist can make about science is that there are far more questions than there are answers. It is the asking of questions that makes science so dynamic; and it is the constant searching for the most probable answers that occupies a great deal of scientists' time and efforts.

The kind of searching that scientists engage in is more often deliberate and systematic than impetuous or haphazard. There is always some sort of sensory basis for any hunch or interpretation. The more supportive sensory evidence is available, the more accurate and generalizable the hunches will be. This chapter centers on three different types or levels of hunches which scientists make, which you can learn to make, and which you can help children learn to make. These are *inferences*, *predictions*, and *hypotheses*—three skills which go beyond the basic skills presented in Chapters 4 and 5.

It may be useful at this point to reconsider Piaget's ideas. The skills of inferring, predicting, and hypothesizing are far more appropriate for concrete-operational children and formal-operational children and adolescents than they are for preoperational children. True, preoperational children can and do frequently make guesses. But a guess is usually illogical and haphazard, with little or no reflective supportive evidence. Inferring, predicting, and hypothesizing are different from guessing. In order to *infer*, one must establish some tentative interpretation regarding cause and effect. To *predict*, one must extend present data into the future—that is, see what present data imply about future events. Hypothesizing is the most sophisticated and mentally demanding of these three skills. To *hypothesize*, one must make generalizations about relationships; these generalizations may or may not be true, but they explain data known to be true. (The distinctions among these three skills should become clearer as you progress through this chapter.) These mental operations are generally beyond the capabilities of preoperational children.

In Chapter 5, four activities of varying degrees of difficulty were presented for each of the basic skills: distinguishing spatial and temporal relationships, classifying, and communicating. I hope you were able to experience the wide range of applicability of these skills and see how each can be used with children at the preoperational, concrete-operational, and formal-operational stages of intellectual development. The applicability of the skills of inferring, predicting, and hypothesizing is not as wide; therefore, only three activities will be presented in this chapter for each skill. This in no way implies that their importance is not as great as that of the skills you have already experienced. On the contrary, after completing this chapter you may decide that these three skills are even more intellectually potent than you initially realized.

Inferring

Suppose you have just walked into a grocery store, and you see one of the checkout clerks begin to cry profusely. Subsequently, a crowd of curious onlookers is attracted to the scene. The following comments are made by some of the spectators:

"She must be in serious pain. I wonder if I should notify the manager or call for an ambulance."

"Perhaps she's just been informed that she's been fired."

"What's that paper she's holding? Why, I bet she won the state lottery!"

"Look! Her cash register is open. Maybe she's been robbed!"

"I believe that last customer said something ugly to her."

"Or maybe he just 'popped the question.' Would you believe that I was working on an assembly line in a candy factory when my husband proposed to me? I was so excited that I ruined the next box of chocolates with my tears."

Each person observed the same clerk crying and formed a different interpretation or explanation of what was perceived. In other words, each person expressed an *inference*. One's inferences are influenced or conditioned by past experiences or observations. Therefore, an inference is more than a guess. On the basis of the limited information available to the onlookers in this example, there's no way of knowing whose inference was correct—if anybody's was. But more important than accuracy is the need to know when one is observing and when one is inferring. That's what the following activities will help you do.

Activities

Puzzling Pictures

A. Introduction   Many of the objects, events, or ideas that you encounter daily elicit the skill of inferring. The example above had to do with an event; this activity has to do with an object. Suppose you are preparing a small plot of ground for a garden and have dug up the object shown in Figure 6-1. Carefully study the details of the different views pictured.

B. Materials   Only the photos in Figure 6-1.

C. Procedures
1.   Formulate some observations about the object pictured. (Of necessity, your observations in this task can be derived only from your sense of sight. But you'll still need to be careful in referring to *only* the physical properties pictured.) Then think about possible inferences—that is, interpretations or explanations of your observations.
2.   Compare the following samples of observations and inferences with those you formulated:

Figure 6-1 Mystery object. (Photographs by Elizabeth Craft.)

Observation A: The object has different shades of color.
Inference A-1: The object is made of different materials.
Inference A-2: The object was painted different colors.
Inference A-3: The aging process caused the object to turn different colors.
Observation B: The surface of the pipe-like part is irregular.
Inference B-1: Parts of the object are rusted.
Inference B-2: The object must be 20 to 30 years old.
Inference B-3: The object probably no longer functions as it was originally intended to do.
Inference B-4: The object might have been buried, or it might have been abandoned and later covered with soil naturally.
Observation C: An enclosed "black box" is next to the gray open-ended pipe-like part.
Inference C-1: The covering of the "black box" is made of plastic or a rubber-like material.
Inference C-2: The "black box" may contain other material, such as oil, wires, springs, or additional smaller "black boxes."
Inference C-3: Power was created inside the "black box" for the other parts of the object.

3. Make a list of at least five to ten additional observations. Formulate several possible inferences for each observation, like the examples above.
4. Then share and compare your results with others. Read one of your statements and determine if someone can identify it correctly as an observation or an inference. Practice this procedure several times until the distinctions between observations and inferences are obvious. Most likely the need for clear communication will also become more evident. It should also become apparent that a variety (or lack) of experiences greatly influences one's inferences. (This is an important realization if you design inferring encounters for children.)
5. Generate some final inferences concerning the use and the age of the object pictured—or any other possible points of interest, like the original owner of

the object or its history. Consider what additional observations and information are needed in order to decide which inferences are more realistic.

D. Extensions   If possible, substitute real objects for pictures. However, select your objects discriminately. It is easier and also more stimulating to formulate inferences about unfamiliar objects than familiar ones.

If you decide to use pictures instead of real objects, try to select ones which raise unanswerable questions or have features that are open to a wide variety of interpretations. A static, placid setting with no discrepancies will be ineffective. Magazines and newspapers often have pictures that can be used to develop the skill of inferring.

The use of pictures can be modified in several ways. One modification entails the use of "picture windows" (Figure 6-2). Mount the picture on the inside of a manila folder. From the front cover of the folder, *selectively* cut out three sides of a frame around a portion of the picture you want to reveal. (The rest of the picture is hidden.) Bend back the uncut part of the frame so that it serves as a hinge for the window. A small piece of masking tape can be attached as a tab for easier opening and closing of the window. Construct as many windows as you wish. You might ask the viewer to open the picture windows in a specified order or randomly. In either case, request observations of the view through each window to be followed by inferences. After all the picture windows have been examined, formulate inferences concerning the composition of the total picture. Decide whether or not the viewer should be permitted to open the folder in order to check the accuracy of the observations and inferences.

Another modification might be entitled "What's Your Version?" Select only pictures which have a brief caption describing their meaning. Keep this description separate from each picture or concealed in an envelope attached to the picture or folder. Ask others to give their own versions (orally or in writing) of each picture. Then share with them the original caption. Analyze which aspects of the descriptions were observations and which were inferences. (If you try this activity with children, don't be too surprised if some refuse to accept the original or objective version and, instead, hold steadfast to their subjective views. Use this opportunity to gain valuable insight into the persistence of egocentricity.)

The use of pictures can also be related to a language arts activity. Select an interesting picture, or have the children select one from your collection, and ask for a written story about the picture. Ask that every phrase, clause, or sentence dealing with inferences be underlined. Constantly help your students to become more aware of the distinctions between observations and inferences and to be conscious of when they are using which.

Mystery Boxes

A. Introduction   You've already encountered the notion of a mystery box in the preceding activity—the "black box" in Figure 6-1. Scientists often encounter mystery boxes both figuratively and physically as they search for answers. For example, the causes of and cures for cancer are to some extent still mystery boxes. The formation of our planet or of the first living cell might for an

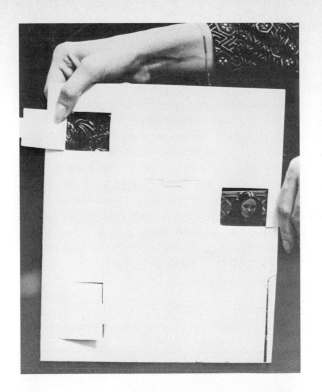

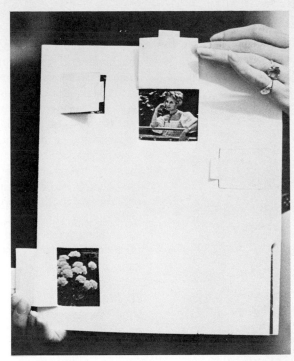

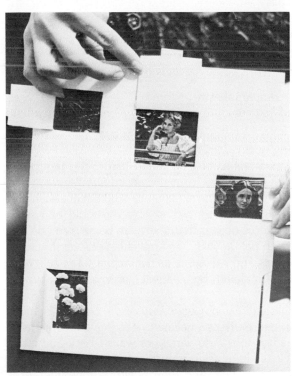

Figure 6-2 "Picture window."
(Photographs by Dale Cornelius.)

indeterminate time constitute mystery boxes (different in size but similar in complexity). Most of us experience mystery boxes every day: the light switch, the water faucet, the food ingested and the body's responses to it, the numerous machines and gadgets used, and on and on and on. Were you to carefully analyze your day, you might come up with several hundred "black boxes." It's easy to take them for granted when they function properly. But when they don't work, we suddenly realize how little we know about them.

Because of the convenience of letting someone else fix our gadgets, or because we've become conditioned to expect them to break down, the investigative skill of inferring can easily become dormant. The feeling that a mystery box is a mystery box and always will be a mystery tends to prevail. However, it is quite likely that the skill of inferring becomes reawakened when you receive a gift-wrapped present. Rarely does one open a wrapped present without first shaking it, pressing the sides, even sniffing at it a few times, speculating on its contents. This activity will give you an opportunity to revivify your skill of inferring about mystery boxes. (Form a small group of participants to carry out the activity.)

B. Materials   (1) At least six or more identical or nearly identical boxes of shoe-box size or smaller; (2) at least twelve or more strong rubber bands; and (3) an assortment of identical or nearly identical *pairs* of objects such as the following:

Golf balls, tennis balls, baseballs
Hairbrushes or toothbrushes
Chains of about seven or more paper clips
Chalkboard erasers
Bars of soap
Ballpoint pens or lead pencils
Potatoes, apples, or rocks of comparable size
Pine cones or loosely wadded balls of aluminum foil

C. Procedures
1. Some advance preparation needs to be completed before the group can participate in this activity. One person from the group should place one object from each pair in the list above in a separate box. Secure each box with at least two rubber bands. The remaining objects should be readily visible. (If there are more pairs of objects than there are boxes, remove and conceal one object from each pair anyway. The matching objects should be placed with the visible objects and can serve as distractors.) The following procedures can now be completed by a small group of three to six participants.
2. First a rule must be established: *Do not open any of the boxes.* Use your senses of hearing, feeling, and perhaps smelling to observe the contents of each box. Try moving the box in a wide variety of ways—back and forth, upside down, sideways, circularly, quickly, slowly, gently.
3. Discuss the observations of each box, and then begin forming inferences

about the contents of each box. How might the movement of a golf ball compare with the movement of a bar of soap? Which objects might make the most similar sounds? Which might make the most different sounds?

4. Without opening the boxes, write your inferred choice of each matching object on a slip of paper and place it next to the appropriate box. Discuss the bases for your choices.

5. Open each box and check the accuracy of your selections. What were the most important clues? What qualities were the most misleading or difficult to interpret?

D. Extensions There are two schools of thought on the question, "To open or not to open the mystery boxes?" In real-life situations, many mystery boxes remain closed for years and years; rarely can inferences be verified simply by removing rubber bands and lids. Many "process purists" would consider opening the boxes to border on scientific sacrilege. And it is true that this can be an excellent opportunity to help children learn to accept and respect the essence of scientific inquiry—that answers, especially "the right answers," are seldom readily available. This is also an opportunity for you to test your patience as the children incessantly demand, "What's in the box?" Recognize the problem, thoughtfully consider the consequences, and make your own decision.

The mystery boxes can be made more challenging by not using matching pairs of concealed and revealed objects. Ask someone to place *any* objects in the boxes—more than one object per box, perhaps—and develop some inferences based on your observations. Don't be concerned with correctly naming the exact object enclosed. Simply try to carefully identify the main characteristics of the object—such as composition (plastic, metal, wood), size, shape, and texture.

Boxes can also be constructed with moving parts, such as arrangements of connected straws or rods extending through small openings. Or you can rig the inside of the box with barriers or a maze. For example, instead of inferring about the object in the box, use either a marble or a small cube, and infer about the arrangement within the box. An interesting assortment of suggestions for mystery boxes can be found in *Developing Children's Thinking through Science* (Anderson et al., 1970, pp. 67–76). Invent your own collection of mystery boxes; you'll be amazed at how much they aid learning.

A Corny Elevator

A. Introduction There are numerous opportunities to infer, of course, other than the use of pictures and mystery boxes. Many times during the day we are confronted with certain results and more or less instinctively begin inferring about the causes. Has anything like the following examples ever happened to you?

Yesterday the tomato plants were watered and all seven of their ripening fruits were examined in anticipation. Today there are only two of the more anemic-looking specimens left. Did squirrels do it? Or rabbits? Could it possibly have been the neighbors?

There's a strange noise in my car motor. Is it the transmission or the clutch? Is the car due for an oil change? Have the gear bands ever been checked? Maybe the spark plugs need to be changed. Is my car another victim of programed obsolescence?

This cake is a flop. Was the recipe followed correctly? Is the baking soda too old? Or was baking powder to be used? Maybe the oven isn't heating correctly.

That garage door was closed this morning. Why is it open now? How was it opened? What should I do?

This flashlight doesn't work. Is the bulb no good? Are the batteries too weak? Is the connection not being made? Maybe it's the switch.

During the past week, what occurrences have you encountered that have caused you to wonder? Besides wondering, did you also formulate inferences? One way to improve the skill of inferring in the ups and downs of daily living is to become more observant. This activity will give you some additional practice. It is a modification of "The Popcorn Elevator" in *Modular Activities Program in Science.**

B. Materials   (1) A small clear glass or jar with capacity of 150 to 200 ml; (2) a source of warm water; (3) some fresh baking soda; (4) some vinegar; (5) five to ten unpopped popcorn kernels; (6) a graduated cylinder or your homemade metric material for measuring liquids (see pages 103–104); (7) a balance; (8) a 5-g mass; (9) a spoon.

C. Procedures
1.   Pour 50 ml of lukewarm water into the glass or jar and then add 5 ml of vinegar. Gently rotate the container to mix the solution.
2.   Then drop three or four of the unpopped kernels into the solution and observe the results for several moments.
3.   Mass about 5 g of the baking soda. Add a small portion of this (about 1 g) to the solution with the kernels and observe the results for a few minutes. Then add another 1 g or 2 g of baking soda to the liquid and observe the effects. Finally, add the remainder of the baking soda.
4.   Try to identify at least five observations of the results. (Perhaps you can make as many as ten to fifteen observations by sharing perceptions.)
5.   Then state as many inferences as you can to explain your observations. Discuss your inferences with others. Which inferences seem to be the most plausible?

D. Extensions   Formulate some inferences concerning the acceleration of the corny elevator. Which inferences are testable? Select at least one of them and test it. Evaluate your results; compare them with others' and determine if there are any foolproof ways of causing the popcorn kernels to move up and down more rapidly. Then test your best inferences on slowing the movement of the

*Adapted from C. F. Berger et al., *Modular Activities Program in Science,* teachers' edition. Copyright © 1974 by Houghton Mifflin Company. Used by permission of the publisher.

kernels. Which inference was proven to be most effective in retarding the movement of the popcorn?

Find a child's windup toy or mechanical toy that no longer moves. Infer the internal composition of the toy and then check out your inferences by dissecting it. Decide what repairs are needed and perform them. What are the effects of your inferences?

Keep a log of any discrepant events that occur during one week. Briefly record your observations and inferences. Whenever possible, test your inferences. Most likely you will discover that the skills of observing and inferring are quite distinct yet mutually supportive.

Additional Considerations on Inferences

One of the greatest benefits of being able to distinguish observing from inferring is a decrease in the tendency to jump to conclusions. Frequently people speak or write as though they were conveying facts or observations when in fact they are making inferences; or at least their listeners or readers are led to interpret their comments as observations. In education, especially, enormous ills can result from such miscues. Children can be branded or categorized for the duration of their schooling as "slow learners" or "troublemakers" as the result of a teacher's "observations" which actually were merely inferences biased by past, limited experiences. Results of educational tests can often have the same effects—even nonjudgmental Piagetian-type interviews. You may have experienced similar phenomena in business transactions, news reporting, advertising, politics, and ordinary social interactions. The over-the-fence chatter or the "just-between-us" divulgences are often replete with fiction taken for facts or inferences mistaken as observations.

The skill of observing enables us to become more aware of our world and maybe even more comfortable with it. The skill of inferring should make us more curious as well as more cautious. Inferences are always tentative, limited, and deserving of scrutiny. They tend to be subjective and conditioned by previous experiences. Seldom should they occur as solitary explanations, since more than one inference is usually possible as well as plausible. With the search for additional observations and information, you may find that inferences can lead to the right answer, or they may lead to dead ends. The scrupulous use of the skill of inferring is absolutely vital to the entire scientific investigative process and, specifically, to the next skill to be examined—predicting.

Predicting

Predicting is one of the most familiar skills. Weather predictions are announced numerous times each day. Anticipated results of sporting events are often publicized. The stock market is a conglomeration of predictions. People often depend seriously on predictions, from the *Farmers' Almanac* to television networks' preelection polls.

What's involved in predicting? What's the difference between an inference and a prediction? Certainly, both are types of hunches. As was mentioned previously, an inference is an interpretation or an explanation based on observations. It is a comparatively simple speculation from immediate observations. (For example, I observe that the sidewalk is wet outside, and I infer that it is raining.) A prediction, on the other hand, can be formulated only after a series of observations and measurements have been made, and their relationships have been determined. On the basis of the analysis of previous data, one can predict or forecast what future observations will be. (For example, I recall observing that the past few days have been very hot and humid. The cloud cover has increased steadily. Today the temperature is decreasing and the clouds are lower. I predict that it will rain.) The more data you have drawn from previous experiences, the more confidently you can predict future occurrences. Is the preceding statement an observation, an inference, or a prediction? You decide, after completing the next activities.

Activities

Starch Search

A. Introduction   It's time again for your physical checkup. But you're a bit anxious about the doctor's reaction to the extra mass you've accumulated since the last visit. On the bases of previous experiences, you can already predict the coming repercussions.

Sure enough—the doctor declares that you must either cut down on starches or start exercising. Later, you realize that you can't distinguish starchy foods from sweet. After some brief research on chemical indicators, you discover that iodine will turn black when dropped on starchy substances. Potatoes and bread are cited as foods high in starch. Your next task is to determine what other foods that you normally eat contain starch.

B. Materials   (1) A small bottle of iodine; (2) a medicine dropper; (3) a small, clear piece of cellophane (about 5 cm by 8 cm); (4) some old newspapers or paper towels; (5) a piece of potato, (6) a piece of bread; (7) a collection of foods normally eaten and readily available—such as cheese, crackers, cookies, cake, cereal, any fruits or vegetables, candy, pretzels, rice, noodles, meats, and so on, according to the dictates of your taste.

C. Procedures
1. Since the previous activities have encouraged distinguishing observations from inferences, it would be wise to first test the "givens," potato and bread. Use the medicine dropper to put two drops of iodine on a piece of potato and then two drops on a piece of bread. Observe the results carefully. If the goldish-brown iodine has changed to a blackish color, you now have a *standard for comparison.*
2. Some difficulty may be encountered when darker materials such as whole wheat bread, pretzels, or brown cereals are tested. Therefore, another standard for comparison would be beneficial. Simply place several drops of

Table 6-1
PREDICTIONS FOR THE STARCH SEARCH

FOOD SAMPLE	STARCH PRESENT	STARCH ABSENT

Table 6-2
RESULTS OF THE STARCH SEARCH

FOOD SAMPLE	ORIGINAL COLOR	COLOR CHANGE	STARCH PRESENT	ACCURACY OF PREDICTION

iodine in the center of the small piece of cellophane. This can serve as your *control sample*. Hold the cellophane with iodine above the piece of food which is being tested and compare the colors of the two samples. If the iodine on the cellophane begins to dry before all foods are tested, make a new standard for comparison. One important caution—Iodine does indeed stain; so use old newspapers or paper towels to cover the work area. (If you try this activity with children, have them wear the smocks they use for art class. Cautioning them against getting iodine on the skin will probably be in vain: many children believe that such stains make them look and feel more like "real scientists." But warn them that iodine should *not* be tasted.)

3. Select at least five to ten food items you wish to test for the presence or absence of starch. List the items in tabular form and specify your predictions. Table 6-1 shows one way of organizing the information.

4. Now conduct your starch search. Try to be consistent with each test; that is, always add the same amount of iodine (for example, two drops) to each food sample. Use initial samples of the bread and potato as well as the pure iodine on the piece of cellophane as standards for comparisons. Record your findings in an orderly fashion, such as that suggested in Table 6-2.

5. Analyze the accuracy of your predictions. Which foods appear to have the greatest starch content? Which results were most surprising? If you outlined a low-starch or starch-free diet for yourself, what would it consist of? Do you think it would be easier to eat less starch or to exercise more?

D. Extensions Perhaps before the diet can be drawn up, additional questions like the following need to be investigated:

Do different kinds of breads, crackers, or cereals contain varying amounts of starch?

Do raw fruits and vegetables contain more or less starch than cooked ones?

Formulate a new set of predictions and then test them.

When introducing the skill of predicting to children, try to select encounters which either stem from their personal experiences (this is a cardinal principle of all learning) or deal with rather regularly occurring events or observations that have a pattern. For example, if the children in a particular classroom were surveyed to determine their favorite desserts or television programs, they could probably predict fairly accurately the preferences of children in another classroom at or near their age level. Or if children observed a lit birthday candle decrease 5 mm in length in 30 seconds and then decrease another 5 mm during the next 30 seconds, they could probably predict the height of the candle after 2 minutes. If children have only a limited repertoire of certain experiences or if the events they are to consider occur irregularly or capriciously, predicting is difficult if not impossible. Frequently, children tend to equate predicting with guessing. During their initial encounters with this skill, you may need to give them considerable guidance so that they can base their speculations about future events on their previous observations and on measured data. Try to focus your initial concerns on the appropriateness of activities related to the skill of predicting rather than on the accuracy or inaccuracy of the children's predictions.

Bouncing Ball Baffles

A. Introduction   We live in a society that seems to be infatuated with bouncing balls. Millions of people spend millions of hours moving and bouncing various kinds of spheres or modified spheres—basketballs, soccer balls, volley balls, bowling balls, footballs, rugby balls, softballs, baseballs, billiard balls, tennis balls, handballs, Ping Pong balls, golf balls, and even marbles—or watching others do it. In all these activities, it is continually necessary to predict the movement of the ball. You can look at the following exercise as good training for ball games, or as simply an interesting and manageable example of predicting. It's a modification of "The Bouncing Ball," Module 48, from *Science: A Process Approach II* (1974, pp. 1–7),* and is intended for a small group.

B. Materials   (1) A meter stick; (2) paper for making a wall chart at least 5 dm to 10 dm wide and 20 dm long; (3) a marking pen which will make heavy, readily visible lines; (4) masking tape; (5) some graph paper; (6) a collection of at least five different balls such as a hollow rubber ball, a solid rubber ball like a handball, a tennis ball, a Ping Pong ball, a golf ball, a basketball, or a volley ball.

C. Procedures
1. Select a fairly open area which is near a wall or closed door and has a hard, uncarpeted floor. The more open the space, the better.
2. Construct a wall chart showing twenty horizontal lines across the width of the paper, spaced 1 dm apart. Number the lines consecutively. The bottom

*Used by permission of the publisher from *Science—A Process Approach II*, American Association for the Advancement of Science, Ginn, Lexington, Mass., 1974.

of the paper should be even with the floor, the first line 1 dm from the floor, and so on. Then tape the chart to the wall or closed door.

3. Remember that in order to predict, you must have a basis of previous observations. Therefore, you'll need to become more familiar with the bounceability of the balls you've selected. Hold a ball about 3 dm away from the 15-dm line on the wall chart. Then release the ball and count the number of bounces the ball makes before the bounces become so rapid that they are uncountable. (The members of the group will need to agree on this point.) Repeat this procedure several times until the number of bounces is agreed upon and the average number of bounces can be determined.

4. Then test the bounceability of three of the other balls. Construct a bar graph which shows the number of countable bounces of the four balls you've observed. If you need a reminder of how to organize the graph, examine Figure 6-3, or review the rules for graphing outlined on pages 142 to 143 of Chapter 5.

5. Interpret your results to determine the order from greatest to least elasticity or resilience of the balls. There is still at least one ball that has not been tested. Predict where it should fit in your graph, and then test your prediction.

6. After these preliminaries, you should be ready to do the more demanding tasks. Use *only* the three bounciest balls. Record the uppermost height of just the first bounce of each ball when released from 5 dm, then from 10 dm, and last from 15 dm. Table 6-3 may be used to organize the data. (Always record the average bounce of at least three tries.)

7. Here, too, a graph will illustrate the results more clearly (see Figure 6-4). You must decide if you want to make separate graphs which show the results for each ball tested. If so, three separate bar graphs will be appropriate. Or, if you wish to incorporate the results of all three balls in one graph, a line graph with color codes or differently slashed lines to identify the different balls could be constructed.

8. Interpret your graph to complete the following statements:

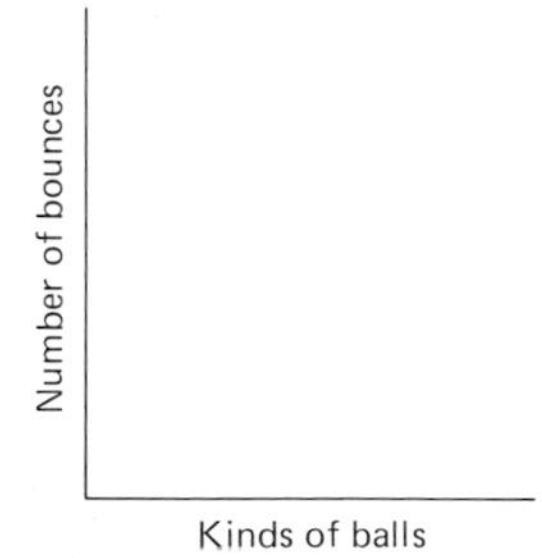

Figure 6-3 Number of countable bounces of various balls.

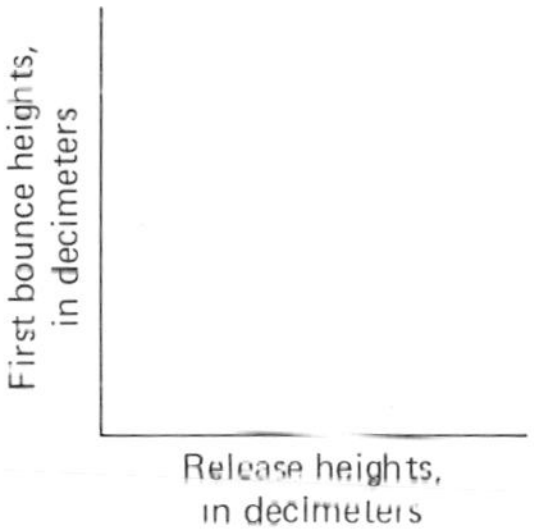

If ball A is released at 7 dm, the *predicted* bounce height will be _____ dm.

If ball C is released at 13 dm, the *predicted* bounce height will be _____ dm.

If ball B is released at 20 dm, the *predicted* bounce height will be _____ dm.

Figure 6-4 Height of first bounce for various balls.

Table 6-3
FIRST-BOUNCE HEIGHTS

RELEASE HEIGHT	BOUNCE HEIGHT		
	BALL A	BALL B	BALL C
5 dm			
10 dm			
15 dm			

Figure 6-5
Two-ruler marble-roller ramp.

9. Test your predictions and evaluate your accuracy. How much has your accuracy improved?

D. Extensions There are numerous variations worth trying:

Select a variety of different release points; formulate your predictions and check out the actual results.

Select a different set of balls or different brands of balls of the same kind.

Change the surface of the floor area.

Try predicting the height of the second bounce.

Generate your own predictions for testing.

You might also try using a setup similar to that suggested in Figure 5-21 for the ruler-ramp marble-roller. Instead of using just one ruler and allowing the marble to roll freely from the ramp, use two rulers end to end as shown in Figure 6-5. The 1-cm ends of the rulers should meet as closely as possible at the midpoint. (Masking tape can be applied to the undersides of the rulers or on the midpoint edges to prevent them from separating.) Identical objects should be used to raise the far ends of the rulers to equal heights or angles. Release a marble from a series of points such as 10 cm, 15 cm, and 20 cm. (Always recheck your procedures and results for consistency.) Record how far the marble rolls up the opposite ramp. Then practice predicting the roll distance for unknown release points. Other variations can be tested, such as:

Changing the size or mass of the marble

Changing the angles of the rulers

Placing a second marble at the midpoint of the rulers and predicting its roll distance when struck by the oncoming released marble

Testing your own predictions

Burned Out

A. Introduction The voice from the radio drones on: "Do you have that run-down, burned-out feeling—as if it's a burden to move your body from place to place—as if all the energy has been drained from your system? If so, try . . ." If you were to look around at the people in your life, you'd probably find a wide range of levels and expenditures of energy. Most likely the same is true of you yourself. We all have our high-gear and our low-gear days. It's often difficult to predict which will occur, partly because predicting is difficult in general and especially difficult when not all the factors involved are within our control or even our consciousness. This activity gives you another opportunity to practice predicting; and it can be envisioned as a simulation of the varying degrees of feeling burnt out. Ideally, this is a three-person activity (unless you are exceptionally dexterous). This basic idea is from "The Suffocating Candle," Module 53 of *Science: A Process Approach II* (1974, pp. 1–11).*

B. Materials (1) A box of small cake candles; (2) some clay (a piece about the size of a golf ball); (3) some matches; (4) a rag; (5) some graph paper; (6) a watch or clock with a second hand; (7) some masking tape; (8) a collection of clear, wide-mouth glass jars of various capacities—for example, a ¹/₂-pt jar (240 ml), 1-qt jar (960 ml), a 1-pt jar (480 ml), a ¹/₂-gal jar (1920 ml).

Warning: Until now, only a few comments about safety have been made. This seems to be an appropriate point for some more direct warnings.

If you will be working in an area with a hard floor, glass breakage can create a dangerous situation. To lessen the danger, place a strip of masking tape around the entire circumference of each jar. This will *not* prevent a jar from breaking if it is accidentally dropped, but it can reduce shattering and splattering.

Also, ask your students to take certain precautions. If possible, require the use of plastic safety goggles whenever there is the least likelihood of accidents with glass. When an open flame is to be used, long hair should always be held back with something like a rubber band. A bucket of water or a blanket should be within reach.

The saying "Better safe than sorry" should not be taken lightly. Children may not be aware of accidents, liability, or lawsuits, but you must be.

C. Procedures

1. Beginning with the smallest container, label the jars A to D. If you were unable to find the exact sizes noted above, use your metric materials for measuring volume and determine the capacities in milliliters. A wide variety of jar sizes should be used.
2. Each participant has a specific job to perform:

The *candle controller* sets up the birthday candle securely in a base of clay and lights the candle. This person also serves as general supervisor.

The *jar judge* inverts the jar over the burning candle, removes the jar after the flame is completely extinguished, and "airs out" the jar before the next test is begun by pushing and pulling the rag in and out of the jar several times. The rag can also be used to wipe away any black carbon which might accumulate on the jar, so that visibility is not reduced.

The *time teller* records the exact number of seconds the candle remains burning after the jar has been inverted over it. This amount will be referred to as the *burning time.*

*Used by permission of the publisher from *Science—A Process Approach II*, American Association for the Advancement of Science, Ginn, Lexington, Mass., 1974.

Table 6-4
BURNING TIMES

JAR	PREDICTED BURNING TIME	TRIALS			AVERAGE BURNING TIME
		1	2	3	

The efforts of the jar judge and the time teller must be carefully synchronized. All participants must also agree on the meaning of *extinction*.

3. Select any one of the four jars and conduct a trial run to coordinate responsibilities and to establish an initial standard for later comparisons and predictions. Determine the burning times for *three* trials with the selected jar. Then calculate and record the *average* burning time (Jar _____; Observed burning time _____).

4. Discuss predictions of the burning times for the other three jars to be tested. Use Table 6-4 to organize your data.

5. Evaluate the accuracy of your predictions. Perhaps you found it more difficult to formulate accurate predictions in this activity than in the previous activities. If that is the case, what might be some of the reasons? In other words, what factors beyond your control might have influenced the results? (If you analyze the potential energy and expended energy of yourself and others, you will also find many factors which are either uncontrollable or unaccountable.)

D. Extensions   Try any or all of the following:

Vary the kind, height, and number of candles used. Predict the results; then test your predictions.

Select some additional jars with different capacities from those previously used. Examine the data already collected and predict the burning times for the new jars. Test; then evaluate your predictions.

Select any three jars which you have already tested and which are of differing sizes. Demonstrate how you can invert each jar over three separate burning candles so that all three candles are extinguished simultaneously. (Yes, it can be done.)

Additional Considerations on Predictions

Although we may sometimes be unaware of it, many of our daily activities are influenced by our predictions. For example, our clothing selections—to wear a coat or a jacket; to take the raincoat or leave it behind—are often based on the weather forecast. Depending on past observations and experiences, we tend to select certain times to go shopping, to take a trip, to plant a garden, or to winterize our dwellings. But we sometimes fail to make predictions. When do you tend to predict? When do you fail to predict and wish that you had predicted? What additional observations would have improved your predicting skill?

Often, in order to predict we must "read between the lines," or "beyond the lines," of observed events. For example, if you knew the burning times of a candle under a 500-ml jar and under a 1000-ml jar, you could also predict the burning times for a 600-ml jar, a 700-ml jar, or any number of variations between 500 ml and 1000 ml. When you know two points or two sets of data, you can predict or *interpolate* other results in between. (Both the bouncing-ball and the rolling-marble activities required interpolations.) You can also predict beyond the data, that is, *extrapolate*. For instance, any prediction of the burning time of a candle in a jar greater than 1000 ml would be an extrapolation. Think about some of your real-life situations. When do you have opportunities to interpolate? To extrapolate?

The greater the spans between the known data, the more risky the predictions. Also, as you've probably already discovered, the more unknown or uncontrolled factors present, the less reliable the predictions. But as was true with inferring, accuracy is not the point of predicting. *Forecasting* is the point, and a forecast is quite different from an inference. Predicting can facilitate preparedness. It will never eliminate anticipation, expectation, anxiety, or the element of surprise. Anticipating future observations is an endless process—made a bit more manageable by the next skill, hypothesizing.

Hypothesizing

The word *hypothesizing* and the phrase *to formulate a hypothesis* are seldom used in casual conversation. You may not even have a very clear idea of hypothesizing but only a vague memory of it as one of the important steps in the "scientific method." What does *to hypothesize* mean? A hypothesis, too, is related to a hunch. But what makes it different from a guess or from an inference or a prediction?

Perhaps some examples might help. Analyze the following sets of statements.

Set I:

1. Green plants require light, moisture, and air in order to live.
2. Fish always swim against the current.
3. Water-soluble substances dissolve faster in hot water than in cold water.
4. The ends of magnets have the greatest magnetic attraction.
5. Short men are more intelligent than tall men.
6. If candles are held at an angle, they will burn faster.

Set II:

1. The violets in the north window bloom almost constantly.
2. That man caught more trout by walking upstream most of the time.
3. Sugar dissolves faster in hot coffee than in iced tea.
4. This magnet can attract a chain of five paper clips.
5. He was the valedictorian, and he's just slightly over 5 feet tall.
6. The rate at which the candle decreased was 5 mm per 30 seconds.

Compare the two sets of statements. The first set deals with generalities which may or may not be true. The second set deals with observed specifics. Set I consists of examples of hypotheses; Set II, of nonexamples. Reexamine the two sets and determine what you think are the key attributes or the distinguishing characteristics of a hypothesis.

Now try to formulate your own examples and nonexamples of hypotheses. After you've made some initial attempts, use the following information to evaluate your efforts. Perhaps some additional assimilation and accommodation may be necessary.

A hypothesis is a general statement about the relationships between or among factors. It's a generalization which covers all similar situations—that is, all things which fall into the same class—even if all instances have not yet been observed. The more related instances that have been observed or inferred, the "stronger" the hypothesis—that is, the more confidence one can have in it. But a hypothesis is *always* tentative; it's always subject to further testing and examination. Additional observations will either provide support or fail to provide it. One negative or contrary observation can disprove a hypothesis. But it is theoretically impossible to prove a hypothesis, since *all* cases—past, present, and future—would have to be tested (Butts and Hall, p. 248, 1975). Hypotheses which have strong support may eventually become theories, and only irrefutable theories eventually become scientific laws; but that is getting beyond the scope of our immediate considerations.

Activities

You've now had opportunities to deal mentally with the concept of a hypothesis. The following activities will provide more concrete experiences in actually using the skill of hypothesizing. These activities have been selected to demonstrate some variety in the formulation of hypotheses—from hypotheses which are readily supportable to those which are more open-ended.

A Balanced Relationship

A. Introduction   There are all kinds of instances of the need to establish and maintain balance. We've already looked at the need for disequilibrium and equilibrium in learning. We also speak of a balanced diet, a balanced checkbook, balanced tires, balance of stereo speakers, and the balance of nature. When we were children, we might have resolved unconsciously the problem of

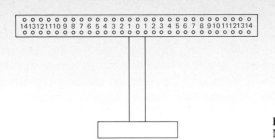

Figure 6-6
Numbered pegboard balance.

balancing our lighter or heavier companions on the teeter-totter. This activity—
without apologies—relates more to the teeter-totter than to the other examples.
You will have an opportunity to observe some balancing phenomena, to
formulate one or more hypotheses, and then to search for supporting evidence.
This is a hybrid activity stemming from Module 78, "Levers," from *Science—A
Process Approach II* (1975, pp. 1–11), and *Senior Balancing* (Elementary
Science Study, 1970, pp. 14–20).

B. Materials (1) A sturdy balance with a pegboard beam; (2) six to ten paper
clips unfolded to serve as hooks for hanging washers; (3) a marble-size amount
of clay; (4) at least twelve or more fairly large washers of identical or
nearly-identical mass. (In case you haven't discovered this already, "identical"
washers are not always of equal masses.) The washers should be heavy enough
so that if just one is hung from the balance beam on a paper-clip hook, the
original position of the beam will be noticeably out of balance. At the same time,
avoid using exceptionally heavy washers unless the balance is quite sturdy.

C. Procedures

1. Remove the balance beam and number the holes in the pegboard as shown
   in Figure 6-6. Either use a numbered strip of tape or mark the numbers
   directly on the beam. Then reassemble the balance and determine if the
   beam can move fairly freely, eventually resting in a horizontal position. If the
   beam is out of balance, apply a small amount of clay to the raised side to
   establish balance.
2. Before you begin to suspend any washers from the pegboard holes, think
   about this question: *In a balanced system, what is the relationship between
   the number of washers hung and the placement of the washers on the
   beam?*
3. Begin testing various combinations of *single* washers hung from one side of
   the beam to balance one paper clip with *two* washers hung on the other side
   of the beam. Record at least four or five combinations of data in the
   following format:

1 washer in position _____ balances 2 washers in position _____.

Try to discover a pattern among the data.

4. Next, vary the number of washers on each side of the beam but still use only *two* positions or paper-clip suspensions. For example, suspend three washers together to balance two washers suspended together; or use four washers to balance one washer. Use any combination of two suspensions. Again, organize your data as suggested below:

_____ washer(s) in position _____ balance(s) _____ washer(s) in position _____.

---

5. Analyze your data; then form a hypothesis about your observations. That is, formulate a general statement of relationships that you think will cover all possible examples. (It might be helpful to reconsider the question noted in item 2.)

6. After you have recorded your hypothesis, test it with several new combinations. Determine if the results support your hypothesis. If the data are not supportive, decide if the hypothesis can be modified or should be dropped. What is your decision?

7. Compare your hypothesis with the following possibilities:

   a. The number of washers on the left side of the beam plus their distance from the center equals the number of washers on the right side of the beam plus their distance from the center.

   b. The mass times the distance on the left side of the beam balances the mass times the distance on the right side of the beam.

   c. The mass divided by the distance on the right side of the beam balances the mass divided by the distance on the left side of the beam.

   Which hypothesis do your data support? Which of the three possibilities is most similar to the hypothesis you formulated?

8. If you are still not sure, test all four hypotheses—yours and the three possibilities presented in item 7. What's your final hypothesis regarding a balanced relationship, at least among washers?

D. Extensions   Examine the two pegboard balances in Figure 6-7. (If you are suspicious, duplicate the arrangements to determine if these are balanced

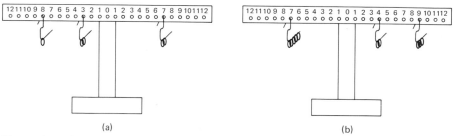

(a)   (b)

Figure 6-7   Combinations of three paper-clip hangers.

relationships.) Construct a hypothesis related to the observations of Figure 6-7. Then collect data to support your hypothesis or to modify it as necessary. Generate several other hypotheses for a variety of different combinations and arrangements. Then evaluate your skill at hypothesizing.

Here is a variation: Instead of beginning with a balanced pegboard beam, place the fulcrum or pivot point nearer to one end—for example, anywhere from five to ten holes from an end. Investigate new arrangements and combinations to achieve a balance. Then formulate new hypotheses and check for supporting evidence.

A pegboard makes an excellent balance beam because the paper-clip hangers cannot slide about easily. However, the more dexterous can use a meter stick arranged as shown in the first example in Figure 4-7, page 102. The extended rod could be a piece of strong coat-hanger wire or the end of a thin screwdriver. Secure the rod with tape or a pile of books. This arrangement allows for a much broader range of combinations and challenges. You may need to use lighter washers or even large paper clips instead of washers.

One very practical suggestion: Develop a set of directions or problem cards to be used by others. Here's one possible example:

The following combinations of washers were hanging on the *left* side of a 33-hole pegboard balance beam:

a. 1 washer in position 12
b. 2 washers in position 9
c. 3 washers in position 6

How many different combinations can you discover to achieve a balanced relationship? What hypothesis will cover all similar arrangements?

Some simple diagrams could also accompany the written directions. If possible, examine the "Problem Cards for *Senior Balancing*" developed by the Elementary Science Study (1970). The set consists of sixty-three 5-in. by 8-in. cards with hundreds of additional ideas on balancing.

Programing Pendulums

A. Introduction In the preceding activity you dealt with some up-and-down phenomena. In this one you will deal with some back-and-forth phenomena by investigating pendulums. You may often have seen examples of pendulums (such as hanging chandeliers, swings in a playground, wrecking balls, and of course pendulums in clocks) without wondering much about them—or about their motion in general. In this activity you will be able to observe more carefully how a pendulum behaves, hypothesize about how you think it should behave, and then check out your ability to program pendulums.

B. Materials (1) At least 1 to 2 m of fishline or heavy nylon thread (cotton string can be used instead, but it tends to stretch more); (2) some paper clips; (3) a piece of clay about the size of a tennis ball; (4) a small *eye screw* (the smaller the better) to which one end of the line can be tied; (5) objects that can be used as masses or *pendulum bobs* (such as plastic pill vials filled with different amounts of sand, or washers); and (6) a watch or clock with a second hand.

Depending on the kind of pendulum support you construct, you may need additional materials. One end of the string should be attached to a raised stable support which allows for the greatest unencumbered swinging movement of the bob attached to the other end of the string. The entire system should be as friction-free as possible. The following are worth considering as pendulum supports:

Securely twist the small eye screw into the middle of a board which is at least 1 m long. The width of the board is not important, but it should be at least 1 cm thick, or strong enough so that it does not bend under slight pressure. With the eye screw pointing downward, position the board on top of the back rests of two tall chairs which are placed about 1 m apart. The seat portions of the chairs should be facing away from the eye screw or in opposite directions to allow for the greatest movement of the pendulum. Two tables could also be used instead of chair backs.

If it's acceptable and convenient to do so, secure the eye screw to the top of a doorway or to an open ceiling beam.

Tighten the eye screw in a small piece of wood and then securely tape this wooden holder to the top of a doorway or to the ceiling.

Investigate the possibility of tying the string to a secure ceiling fixture.

See Figure 6-8: With a single-edge razor blade, make a slit about 2 cm long through the middle of one end of a tongue depressor. Gently wedge about 5 to 10 cm of the string through the slit without splitting the wood. Wrap this short length of string around the unslit part of the depressor, or tape it, to prevent slippage when the bob swings from the opposite end. Then securely tape the unslit half of the tongue depressor to a table or countertop so that the slit end with the hanging string extends outward at least 6 cm. Or tape the tongue depressor to the top of a doorway or any other high, horizontal open surface.

You should be able to implement at least one of these suggestions, or glean from them an original method that can be used in your particular setting.

C. Procedures

1. After selecting and establishing a suitable pendulum support and attaching one end of the string, decide on (a) the length of string you wish to use (try to use at least 40 cm) and (b) the bob to be attached to the free end of the string. Then spend some time investigating the movement of the bob when released to swing from various positions.
2. Determine how many complete swings or back-and-forth trips the pendulum makes in 30 seconds. A complete swing is also called a *vibration* or a *period*. Keep a record of the *frequency* of your present pendulum, that is,

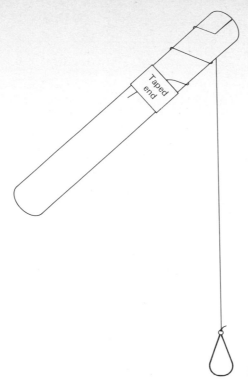

Figure 6-8   Tongue-depressor pendulum.

the number of vibrations per 30 seconds. Use this information as a basis for later comparison (vibrations/30 seconds = frequency).

3.   The purpose of this activity is to program pendulums. Therefore, you need to think about two important questions: (a) *What factors will cause the pendulum to swing faster?* (b) *What factors will cause the pendulum to swing slower?* These two questions lead to the basic question—namely, (c) *What factors influence the frequency of a pendulum?*

4.   Examine your pendulum and its behavior again. What factors might you change which will alter the frequency? Which of the following might cause the most significant effect on the swing of the pendulum?

Kind of support used

Amount of force put into the initial sendoff of the bob

Distance from the resting point where the bob is held before it is freely released (another name for this factor is *displacement*)

Length of the string

Kind of string used

Mass of the bob

Shape of the bob

Other factors not listed here

Table 6-5
PENDULUM TESTING

Hypothesis 1: Results of testing:	
	_____ Frequency
Hypothesis 2:	
Etc.	

5. List a variety of hypotheses related to any of the factors you wish to consider. One possible format that may be used is the "If . . . , then . . ." proposition. For example: "If the support is attached to a structure at a higher elevation, then the pendulum will move faster." Another format is: "The greater the angle of displacement, the slower the frequency."

6. Examine your list of hypotheses and select at least *three* which you think are most worth testing. Try to test your hypotheses systematically—that is, vary only one factor at a time. For example, if you wish to test the effect of the mass of the bob, keep all other factors the same and vary only the mass of the bob. (One way to accomplish this is to tie a paper clip to the end of the string and then attach a marble-size piece of clay. Record the results and then systematically add more clay.) Use the data collected in item 2—the original frequency you recorded—as a basis for all comparisons. Table 6-5 shows one way of organizing your investigations.

7. Examine your data and decide which factor has the greatest influence on the frequency of a pendulum. Evaluate your newly developed prowess in programing pendulums to swing swiftly or slowly.

D. Extensions Describing results of tests by the narrative method is not always the easiest or clearest means of communicating. Often, a graph of the data is much more effective and efficient. The entire data-collecting process also tends to become more systematic when the results are to be presented in a graph. Set up another pendulum and investigate the varied effects of the following: (1) the mass of the bob, (2) the displacement or the release of the bob from its resting midpoint, and (3) the length of the string. Determine the frequency and construct a separate graph for each set of data. One sample is shown in Figure 6-9. For all three graphs, the result is the number of swings per specified time period or the frequency either per minute, per half-minute, or whatever. These data should be placed on the vertical axis. The manipulated factors, or the conditions *you* are changing—the mass of the bob, the displacement, or the length of the string—should be placed on the horizontal axis. Figure 6-10 illustrates one way of measuring the displacement. (A protractor could also be positioned behind the pendulum support with the 90° angle aligned with the pendulum at rest.) After completing the three graphs, interpret the results. How well do these conclusions support your previous hypotheses?

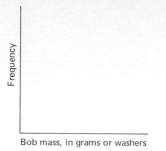

Bob mass, in grams or washers

Figure 6-9 Data from testing pendulum.

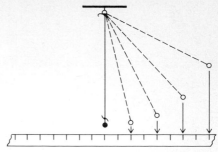

Figure 6-10 Measuring displacement.

If you wish to make further investigations or hypotheses with a pendulum, examine the Teacher's Guide for *Pendulums*, one of the Elementary Science Study units, for additional ideas (1969, pp. 1–33). The suggested activities include effecting a variety of circular motions of the bob; planning and causing the bob to knock down a specific arrangement of inverted standing golf tees; programing coupled pendulums (two separate pendulums whose strings are connected with a thin dowel rod or plastic straw); creating designs with pendulums whose bobs contain salt, sand, or paint; and many more.

Boat Building for Cargo Contracts

A. Introduction   When applying for a job or considering a task to be done, we usually pause to see if we have the necessary skills. If we lack the skills but the job or task is important to us, we'll work toward developing what is needed. Suppose you were given the opportunity to join the ranks of the great shipping magnates. Certainly, it would be important for you to know something about boats, their design and floatability; and about cargoes, their mass and placement. Suppose, too, that instead of purchasing boats from others, you could build your own. This activity gives you the opportunity to formulate and test hypotheses on building boats to float certain cargoes.

B. Materials   (1) A bucket or dishpan (or any other comparable container), three-fourths filled with lukewarm water; (2) a piece of clay or plasticine about the size of a golf ball, with a mass of about 50 g (use *only* an oil-base clay, as potter's clay or "play dough" will eventually dissolve in water); (3) a 15-cm-square piece of aluminum foil; (4) a metric ruler; (5) a graduated cylinder; (6) thirty to forty pennies or uniform small washers of comparable mass (approximately 3 g each); (7) ten or more marbles (ordinary marbles have a mass of about 5 g); (8) some paper towels; and (9) a collection of "test boats" similar to the following:

Plastic lid of an 8-oz margarine tub or a 1-lb coffee can
Lid of a cottage cheese container
Styrofoam cup cut down so that the sides are 2 to 3 cm deep
30-ml medicine cup, or a shot glass
Metal jar lid with a diameter of 5 to 10 cm

Table 6-6
DATA ON TEST BOATS

TEST BOAT	PHYSICAL PROPERTIES	CARGO CAPACITY		COMMENTS
		MARBLES	PENNIES	

C. Procedures

1. Before you will be able to hypothesize about boat building and cargo capacities, you'll need to know more about each of the "test boats" you've selected. For example, what are the boats' physical properties, such as size, mass, capacity, material, and shape? When placed in the container of water, how many pennies can each boat hold before sinking or capsizing? How many marbles? What special discoveries can you make about each "test boat"? Keep a record of your findings, as suggested in Table 6-6.

2. Analyze your data to identify the best penny carrier and the best marble carrier. Which boat might handle the largest cargo of *combined* pennies and marbles? Test your ideas and record your results.

3. On the basis of the observations you have made so far, what factors do you think are most important as regards cargo capacity? Again, keep a record of your ideas. (Many of us have a tendency to think in circles, but we generally write in straight lines.)

4. By now you've probably learned a great deal about the test boats. But suppose that your first assignment has to do with boats to be constructed of clay. What must you do to the clay ball in order to get it to float? Test your ideas, evaluate the results, and make any modifications necessary to achieve a seaworthy clay boat.

5. Suppose you have contracted to build a clay boat which must be able to carry at least a 40-g cargo consisting of both pennies and marbles. Plan and test a hypothesis for a clay boat that will meet this specification. (The "If . . . , then . . ." format might again be helpful. For instance, "If the clay boat has the dimensions x by y by z, it will float 40 g." Or, "If the 40-g cargo is arranged thus, the boat will float.") Evaluate your results and make any necessary changes in your hypothesis. (See Table 6-7.)

6. As any old shipping magnate knows, success breeds success. Now you are

Table 6-7
CLAY-BOAT HYPOTHESIS

Hypothesis: Results:

Table 6-8
ALUMINUM-FOIL-BOAT HYPOTHESIS

Hypothesis:

Results:

Table 6-9
BEST-BOAT HYPOTHESIS

Hypothesis:

Results:

ready for the next cargo contract. This one calls for a boat constructed from the 15-cm-square piece of aluminum foil which must be able to sail in rainy areas. Besides floating a 50-g cargo of pennies and marbles, it must also be able to hold 30 ml of water. Formulate and test a hypothesis for an aluminum-foil boat to meet these specifications. (See Table 6-8.)

7. If you've mastered the last task, you are now ready to join the ranks of the first-class hypothesizers. On the basis of all the preceding experiences and data analyses, formulate and test a hypothesis for a boat constructed from any material that will carry the greatest amount of cargo for the longest period of time. The only restriction is that the boat must be originally constructed by you. In other words, you may not use something like a factory-made styrofoam cup and simply launch it with a full cargo. You may, however, share your ideas and results with other boat builders. Use Table 6-9 to record your efforts, and decide whose boat is best.

D. Extensions   One of the difficulties that might be encountered in evaluating the boats' performances is the lack of some objective norm. Suppose the data shown in Table 6-10 had been collected. They have been listed from least to greatest carrying capacities, but how much better is one boat from another? What is needed is a *boat effectiveness index*, or BEI. This BEI is a comparative ratio obtained by dividing the boat's cargo mass by its boat mass. Then the data on the four boats could be compared more easily, as shown in Table 6-11. The larger the BEI, the better the boat. Hypothesize on various masses, designs, and cargo capacities of clay boats. Test your hypotheses to determine the best boat by comparing their BEIs.

You might also hypothesize about the effect of using soapy water or salt water. Or you could reverse the strategies you have been using—instead of

Table 6-10

BOAT	BOAT MASS, g	CARGO MASS, g
A	50	10
B	40	15
C	5	50
D	8	90

Table 6-11

BOAT	CARGO MASS, g	BOAT MASS, g	BEI
A	10	50	0.2
B	15	40	0.375
C	50	5	10
D	90	8	11.25

building the best floating carriers, hypothesize about how to convert plastic cups into the best submarine carriers.

Additional Considerations on Hypotheses

The ability to know when we are inferring as opposed to observing can lessen the danger of jumping to conclusions. The skill of predicting can lead to better preparedness for future events or observations—to a more rational approach to planning in general. But what about the skill of hypothesizing? Is it all that important, or is it used much in our lives?

Have you ever found yourself either making or believing statements like the following?

People always act stranger when the moon is full.
Frogs (or is it toads?) cause warts.
Hot-water pipes freeze sooner than cold-water pipes.
Vitamin C prevents colds.
Behavioral problems increase when "slow" children work together.
Catholics always have large families.
Soft drugs lead to hard drugs.
Kids from "the other side of the tracks" will never amount to anything.

You could probably add to this list of hypotheses. The point is that more hypothesizing goes on every day than we might realize. But what often fails to occur is the recognition that such generalizations about relationships need empirical testing. A hypothesis is *not* a law or a proven fact.

We need to refer more often to the seven values of science mentioned in Chapter 1, especially the search for data, the demand for verification, the respect for logic, and the consideration of premises. Were we to apply these values more consciously, there would be fewer inferences mistaken as observations and fewer hypotheses accepted as truths. However, such clarifications will not be made until people become dissatisfied with their present approaches to problems or experiences. The skill of hypothesizing itself will not transform you (or your students) from an illogical to a logical thinker—or from the concrete-operational level of development to the formal-operational level, where hypothetical-deductive reasoning is automatic. But more frequent encounters which elicit this skill will certainly contribute to the gradual attainment of such goals.

Inferring, predicting, and hypothesizing are means for seeking explanations or causes and for identifying relationships, interactions, and patterns. But, perhaps most important, these skills can help you develop one of the most important attitudes toward science with children, that is, *respect for tentativeness.*

Encounters

1. This may be a difficult question to answer. It is deliberately intended to require some soul searching. The three words *inferring*, *predicting*, and *hypothesizing* probably were not entirely new to you. But in what ways (if any) has your initial understanding of these concepts changed as a result of experiencing them as thinking skills in the activities in this chapter?

2. Analyze your conversations, thoughts, and actions during the past several days. List some of the inferences, predictions, and hypotheses you've made. Which of the apparently factual statements you made were actually inferences?

3. Refer to any of the hypothesizing activities you completed. Analyze the thought processes you used in them. Then review the hypothetical-deductive operations described in Chapter 2, page 37. What similarities and differences can you identify?

4. Examine some elementary science textbooks or curricular materials for examples of lessons which (a) introduce the concepts of inferring, predicting, and hypothesizing to children and (b) employ these skills in subequent activities. Evaluate the appropriateness of the approaches used and the concepts to be learned—in terms of either the three skills themselves or the scientific facts to be learned while using the skills. In your judgment, was there very little, some, or great potential for false accommodation? What's the basis for your decision?

5. Develop your own inferring, predicting, and hypothesizing activities for children. Draw from any of the previous activities (especially from the Extensions), the sources listed in the References or the Suggested Readings, or any other science curriculum materials. If possible, test your activities with children and modify them if necessary. (You might discover that even a lesson plan is a prediction.) Share your results with your colleagues.

6. Reread the last paragraph of this chapter. In what ways do you agree or disagree with the importance of "respect for tentativeness" as, perhaps, the prime derivative from the three skills presented?

7. Your choice.

References

Butts, David P., and Gene E. Hall: *Children and Science: The Process of Teaching and Learning,* Prentice-Hall, Englewood Cliffs, N. J., 1975.

Anderson, Ronald D., Alfred DiVito, Odvard Egil Dyrli, Maurice Kellogg, Leonard Kochendorfer, and James Weigand: *Developing Children's Thinking through Science,* Prentice-Hall, Englewood Cliffs, N. J., 1970.

Berger, Carl F., Glenn D. Berkheimer, L. E. Lewis, Harold T. Newberger, and Elizabeth A. Wood: *Modular Activities Program in Science,* teacher's annotated edition, Houghton Mifflin, Boston, 1974.

Elementary Science Study: *Pendulums,* Webster Division, McGraw-Hill, St. Louis, Mo., 1969.

————: *Senior Balancing* and "Problem Cards for *Senior Balancing,*" Webster Division, McGraw-Hill, St. Louis, Mo., 1970.

Science—A Process Approach II, Module 48, "The Bouncing Ball," Ginn, Lexington, Mass. (American Association for the Advancement of Science), 1974.

Science—A Process Approach II, Module 53, "The Suffocating Candle," Ginn, Lexington, Mass. (American Association for the Advancement of Science), 1974.

Science—A Process Approach II, Module 78, "Levers," Ginn, Lexington, Mass. (American Association for the Advancement of Science), 1975.

Suggested Readings

Butts, David P., and Gene E. Hall: *Children and Science: The Process of Teaching and Learning,* Prentice-Hall, Englewood Cliffs, N. J., 1975. (Chapter 6, "Explaining What You Observe," and Chapter 11, "Generalizing," are especially appropriate. Additional ideas on inferring, predicting, and hypothesizing are presented.)

Commentary for Teachers: Science—A Process Approach, Ginn, Lexington, Mass. (American Association for the Advancement of Science), 1970. (This teacher's resource book contains specific sections with numerous activities on the process skills presented in this chapter.)

Elementary Science Study (ESS): *Batteries and Bulbs; Drops, Streams, and Containers; Ice Cubes; Microgardening; Mobiles; Sink or Float; Structures; Tracks; Where Is the Moon?* Webster Division, McGraw-Hill, St. Louis, Mo. (These units provide excellent experiences with minimal requirements for specialized equipment for developing the skills of inferring, predicting, and hypothesizing.)

Hungerford, Harold R., and Audrey N. Tomera: *Science in the Elementary School,* Stipes, Champaign, Ill., 1977. (In a workbook format, the authors present a variety of activities on inferring and hypothesizing. See pages 50 to 63, plus Part V in general for a broader view of process models.)

Science—A Process Approach II, Ginn, Lexington, Mass., (American Association for the Advancement of Science). [The following modules are worth examining for additional activities on the three process skills of this chapter:

Module 33, "What's Inside?" (inferring)
Module 38, "Using Graphs" (predicting)
Module 40, "How Certain Can You Be?" (inferring)
Module 44, "Surveying Opinion" (predicting)
Module 50, "The Clean-Up Campaign" (predicting)
Module 58, "Liquids and Tissue" (inferring)
Module 61, "Circuit Boards" (inferring)
Module 70, "Conductors and Nonconductors" (hypothesizing)
Module 73, "Solutions" (hypothesizing)
Module 79, "Animal Behavior" (hypothesizing)

Modules 33 to 58 are appropriate for grades 2 and 3; Modules 58 to 79 are appropriate for the intermediate grades.]

CHAPTER 7
INTEGRATING THE SKILLS

Very little of the learning that occurs in daily life or within the confines of the classroom can be departmentalized, encapsulated, or dosed out in pure forms. It's difficult to conceptualize "pure" science, mathematics, language arts, or social studies. (The last two, especially, would be real challenges to the imagination.) Each of these curricular areas overlaps and interrelates with the others. The same is true of the thinking skills of science: they too are related to other curricular areas.

Observing is as basic to the language arts, mathematics, or any other subject as it is to science. Measuring is certainly used in more than the study of mathematics. Communicating and classifying are related to all types of learning. If you analyze the typical elementary school curriculum, you will eventually discover that not one of the thinking operations—the process skills—presented thus far is confined solely to science.

The process skills also overlap each other. The activities on inferring, predicting, and hypothesizing in Chapter 6 incorporated the skills of observing, measuring, distinguishing spatial and temporal relationships, classifying, and communicating. In fact, all the activities in Part Two are identified with specific process skills mainly for the sake of emphasis. The actual name by which any of these skills is called, its degree of difficulty, and the grade level or age level for which it is an appropriate part of the elementary science curriculum are subject to interpretation by science curriculum writers and publishers. How many of the process skills are presented, and how they are used, also varies from program to program.

On the basis of my own experiences, I believe that observing, measuring, distinguishing spatial and temporal relationships, classifying, and communicating are the most appropriate process skills for children at the preoperational and concrete-operational levels of intellectual development. Generally, children who are strong conservers and who are approaching the formal-operational stage (that is, those in transition between the concrete-operational and formal-operational stages) can use the skills of inferring, predicting, and hypothesizing in new encounters.

The skills to be examined in this chapter—namely, *controlling variables*, *interpreting data*, and *defining operationally*—incorporate most or all of the skills that have already been presented. In addition, they require the ability to analyze and synthesize a greater amount of information which is recorded graphically, stored mentally, or both. To use these skills, one must be able to think about thinking. Most often, the concrete-operational child is unable to perform these intellectual maneuvers consistently or without assistance. Yet, as was explained in connection with hypothesizing, learners greatly need experiences with these integrated skills in order to make the transition from concrete-operational to formal-operational thinking.

The more adept you become at using these mental operations, the better you will be able to guide students in their use. After you have completed the activities in this chapter and, perhaps, had the opportunity to try them or similar lessons with children, decide for yourself about the appropriateness of these integrated skills for various ages and grades—as well as their overall importance.

Controlling Variables

We are constantly being bombarded by advertising urging us to buy new or improved products. A great many of the advertisements use wording such as "the best," "the fastest," and "the longest-lasting—on the basis of *scientific tests." Scientifically tested* has become one of our most frequently used (and abused) phrases. The reasoning which advertisers expect consumers to use is, "It's scientifically tested; therefore, it's got to be good." But what does *scientifically tested* mean? How valid or accurate are the procedures and the results of the testing?

Suppose two people in an advertisement are comparing the strength of two brands of paper towels by rubbing the dampened towels on a counter covered with scouring powder. At the conclusion of the test, they lift up their paper towels: one looks none the worse for wear; the other is in shreds. The results are obvious; the procedures, however, need closer scrutiny. More specifically, the *variables* need to be investigated—that is, the factors that influence the results. (Another meaning for *variable* is "any factor that may vary or might be varied.") In this example, the following variables need to be considered:

The rubbing actions of the two people. Was the pressure applied the same in each case? Were the speed of the motions and the number of motions equal?

The amounts of abrasive on the surface. Were both the same in terms of quantity and quality? Were both areas of the surface initially the same?

The amount of water used to dampen the towels. Were equal amounts used? Was the water "pure" in both cases?

The original size of the paper towels. Were both the same size? The same thickness?

There are other, less obvious, variables that might have effected the results, but those listed above will suffice for now. In order to test something accurately, the variables must first be identified and then regulated or controlled during the actual testing. Only in this way can it be determined which factor had the greatest effect on the results. If only one variable is altered while the others are kept the same or equal, the results can be attributed to the variable that was changed. Otherwise, one guess is as good as the next; there's no way of knowing.

Types of Variables

The idea of variables becomes clearer when you can recognize four major kinds of variables:

1. *Responding variable or dependent variable.* A responding variable is the result or measured change that occurs because of the influences of other factors. For example, in "Programing Pendulums" (pages 169 to 173), the responding variable is the recorded number of complete swings per minute or per 30 seconds—that is, the frequency. In the example of the paper-towel demonstration, the responding variable is the condition of the towel after scouring powder is rubbed in.
2. *Manipulated variable or independent variable.* This is the factor that is

systematically altered, changed, or modified and subsequently may influence the responding variable. In the first suggestion under "Extensions" in "Programing Pendulums" (page 172), you were asked to systematically vary the mass of the pendulum bob and determine the effects on the frequency. You were also asked to vary the release positions or displacements, and then to vary the length of the string. These are examples of three manipulated variables to be tested separately. The manipulated variable in the paper-towel demonstration is the brand.

3. *Controlled variable or variable held constant.* This is the factor that is kept the same while one variable is changed or manipulated and another variable is responding. For example, when the manipulated variable in "Programing Pendulums" is the length of the string, the controlled variables are (a) the kind of string, (b) the pendulum support, (c) the mass of the bob, (d) the displacement, and (e) the free release of the bob from the displacement point. In other words, all controlled variables are kept the same each time the length of the string is varied. The questions raised above about the equality of the pressure and motions applied, the amounts of abrasive and water, and the size of towels used all dealt with variables held constant. Other variables, such as atmospheric pressure, time of day, humidity, air currents, lighting, temperature, and contaminants, should also be controlled or kept the same during testing.

4. *Uncontrolled variable.* This is an influencing factor which is knowingly or unknowingly ignored in the belief or hope that its effects are insignificant. These variables are beyond the investigator's control; examples are the possibility of human error and faulty equipment. (These factors can often be controlled only under the most rigorous laboratory conditions. You should, however, be aware of their presence and potential influence.)

Activities

"Scientifically Tested"—See for Yourself

A. Introduction   You can use the skill of controlling variables to conduct your own testing of products. For example, what is your reaction to the advertisement shown in Figure 7-1 (page 184)? Does such a towel exist in today's market? What towel might be its nearest competitor? Have Tonnage paper towels really been scientifically tested?

B. Materials   (1) At least five sheets of four or more different brands of paper towels; (2) a wide-mouthed metal container (such as a 3-lb coffee can or a metal bowl with a comparable diameter); (3) a strong rubber band that fits firmly around the metal container yet is easy to slip over the opening; (4) a cake pan or any four-sided tray at least 30 cm long, half filled with water; (5) a flat plastic lid similar to that from a margarine tub; and (6) an assortment of gram masses which amount to at least 1000 g. (If you have difficulty collecting enough gram masses, use cans of food that are calibrated in grams, along with coins or washers of known masses.)

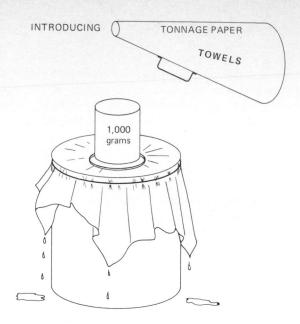

INTRODUCING TONNAGE PAPER TOWELS

The TOUGH TOWEL for TOUGH TOIL!
SCIENTIFICALLY TESTED
to hold 1,000 grams -- dripping wet!
No strings attached
(in the towel)

Figure 7-1
Hypothetical advertisement for analysis.

C. Procedures

1. Since Tonnage paper towels have not yet appeared in stores, select three to five other brands of paper towels to test. After analyzing the results, you will be able to determine more accurately the nearest competitor to Tonnage towels, as well as the probable validity of the test described in the advertisement. But first, try to identify the manipulated variable and the responding variable that are involved in the test.
2. Check your ideas against the following:
    a. The manipulated variable is the brand of paper towel being tested.
    b. The responding variable is the amount of mass supported by the paper towel before it tears.
3. The following variables will need to be controlled:
    a. The "dripping wet" state of the towel. Place the paper towel in the pan of water. After it is completely soaked, lift it above the water pan by holding two adjacent corners. Then let it drain over the pan while being held in this position for 20 seconds.
    b. The placement and attachment of the paper towel over the opening of the metal container. Try to center the towel over the container opening without sliding or pressing it. Stretch the rubber band over the top of the container so that the paper towel is held in place.
    c. The placement of the masses upon the paper towel. Gently place the flat plastic lid in the center of the towel. Then gently place the masses on

Table 7-1
RESULTS OF PAPER-TOWEL TESTING

BRAND	AMOUNT OF MASS SUPPORTED				RANK ORDER OF STRENGTH
	TRIAL 1	TRIAL 2	TRIAL 3	AVERAGE	
A					
B					
Etc.					

the plastic lid. With this procedure, the mass will always be distributed evenly over the same surface area.

4. Keep a record of your results, as suggested in Table 7-1, and evaluate the strengths of the towels from first to last place.

5. How do your results compare with those advertised in Figure 7-1? Were any of the towels you tested able to hold 1000 g? Which supported the most mass? What variables were not accounted for in either your tests or the test of Tonnage towels? In case you need a hint, think about the variable *time*, or the number of seconds a suspended towel could support a certain mass. On the basis of what you've seen for yourself, would you consider the advertisement for Tonnage towels to be false advertising or a claim from incomplete evidence?

D. Extensions If you're uncertain about the answer to the last question above, recheck your procedures—but this time determine the number of seconds or minutes each paper towel can support the greatest mass. Which towel was able to support the greatest mass for the longest period of time? In what way does the inclusion of the time variable change your initial rank order of strength?

Design a test to analyze the absorbing power of paper towels. Identify the manipulated variable and the responding variable. What variables will you try to control? How does the rank order of absorbing power compare with your data on the strength of the paper towels?

Investigate the unit "Consumer Research—Product Testing" from *Unified Science and Mathematics for Elementary Schools* (1976). This unit contains an abundance of additional ideas for testing products such as glue, tape, batteries, sponges, soap, cleansers, plastic wraps, pens, pencils, seeds, and even peanut butter and lollipops. The entire area of consumer research has enormous potential for activities which emphasize controlling variables. An additional benefit might be wiser purchases of "scientifically tested" products.

Crawling Spool Marathon

A. Introduction Imagine that the energy crisis has reached its climax. All sporting events which consume fossil fuels—automobile racing, motorcycle racing, and motorboat racing, and so on—have been prohibited. But all is not lost. Someone puts together a crawling spool system (like the one in "Do-

It-Yourself Picture Power," page 138) and challenges all comers. Thus begins a new era of racing. Eventually, a Board of Controllers is established to regulate the Crawling Spool Marathon (CSM). The following classes or divisions of racing spools are established:

Super spools, ranging in height from 3.5 cm to 6.5 cm and in diameter from 3 cm to 4 cm

Moderate (or "mod") spools, ranging in height from 2 cm to less than 3.5 cm and in diameter from 2 cm to under 3 cm

Peewee spools, less than 2 cm in height and less than 2 cm in diameter

The Board of Controllers also establishes the following rules:

The length of the drag rods may vary from 20 cm to 4 cm.

The length of the unstretched rubber band may never be more than 1 cm longer than the height of the spool.

The flanges of the spool or the outer rims may be either smooth or notched.

On the entry blank, the controlled variables must be identified so that spoolers (that is, spool racers or owners) will compete with each other only under comparable conditions. The responding variable will always be the metric distance traveled by the crawling spool.

Before competing in the CSM, a spooler may conduct any number of trial runs to test the manipulated variables.

B. Materials   Refer again to "Do-It-Yourself Picture Power" (Chapter 5, page 138). This time, assemble a greater assortment of all the components pictured, and have a meter stick or metric tape for measuring the distances.

C. Procedures
1. If you have dismantled your original crawling spool from Chapter 5, assemble another one. Test its performance several times, and keep a record of the average distance traveled and the conditions under which the test runs were made. (See Table 7-2.)
2. The data recorded in Table 7-2 can now serve as a basis of comparison for

Table 7-2
TEST SPOOL DATA

Size and class of spool:
Description of flanges:
Length of drag rod:
Number of rubber band turns:
Type of surface traveled on:
Any other pertinent factors:
Average distance traveled (responding variable of test spool):

Table 7-3
RESULTS OF VARIOUS TEST RUNS

Manipulated variable 1: Responding variable 1:
Manipulated variable 2: Responding variable 2:
Etc.

the next tests. Decide what variables you want to control, and keep a record of them. Then decide what variables you want to test or manipulate. (Some of the more likely possibilities are size of spool, length of drag rod, number of turns of the rubber band, and type of surface of the spool rims or flanges.) Systematically test—that is, test *separately*—each manipulated variable. Keep a record of the results of each test run, as suggested in Table 7-3.

3. After you have exhausted your ingenuity at testing, compare all results (responding variables) with the initial results of the test-spool trial run. Which spool system produced the best results?

4. You are almost ready to enter the CSM. But first the entry blank must be completed. (See Table 7-4.)

5. Discuss with other spoolers the controlled variables, the manipulated variables for each test run, and the responding variable recorded after each trial. In accordance with the established spool classes, plan and conduct a CSM.

6. Evaluate the performance of your spool and its competitors. How closely did the actual responding variable match your expected travel distance? Which responding variable was a CSM record? What words of wisdom on variables would you pass on to future spoolers?

D. Extension Perhaps you've discovered that there is already a shortage of wooden spools. If wooden spools eventually become extinct, the basic idea of this activity can still be salvaged: paper airplanes or kites could be substituted. If you were one of the Controllers of the Paper Airplane (or Kite) Program, what rules would you think should be established? If you were entering a paper-airplane or kite competition, what variables would you need to consider? What would your best paper-plane or kite entry look like? What might be its best long-distance flight?

Table 7-4
CSM ENTRY BLANK

Controlled variables: Expected travel distance:

Variations on a Sprout Search

A. Introduction   Two farmers, Abe and Zack, were having an argument over the effects of a fertilizer solution that was supposed to improve the germination time of seeds. They finally realized that the only way to resolve the argument was to actually test different amounts of the fertilizer in water solutions.

Abe decided that he would plant some corn near the chicken yard, some wheat near the barn, and some soybeans near the storage shed. He would water the plots every day with some fertilizer solution and keep a record of what happened. It's as simple as that, according to Abe.

Zack thought it might not be so simple—that if Abe's plan were used, they would never know exactly what factor caused the seeds to sprout. (Both farmers needed to read Module 71, "Seeds and Soap," of *Science—A Process Approach*, 1975, pp. 1–11.) They then began to think about different variables and how they might be controlled: see Table 7-5.

They still had to decide how to handle the manipulated variable—that is, different concentrations of the fertilizer in water for different seed containers. But by now both realized that the search for causes and effects wasn't so simple.

In this activity, you will be able to realize this for yourself. Select a problem you wish to investigate which deals with the germination or growth of seeds.

B. Materials   Your selection of materials will depend on the problem you decide to investigate. However, if you wish to use seeds that normally have a relatively fast germination rate, use radish, grass, or mung bean seeds. Commercial potting soil or vermiculite (a cork-like mixture of absorbing particles) can be used if you wish to control the soil. If the seeds are planted in identical containers, provide for water drainage and try to avoid overwatering.

Table 7-5

VARIABLE	MEANS OF CONTROL
1. Kind of seed	1. Use only one kind of seed.
2. Number of seeds	2. Use the same number of seeds for each test.
3. Soil	3. Plant seeds in identical containers holding the same kind and same amounts of soil.
4. Planting depth	4. Use a marked pencil for making a hole so that each seed can be planted at the same depth.
5. Watering conditions	5. Always apply the same volume of water to all seeds at the same time of day.
6. Environmental factors such as light, temperature, humidity, and wind	6. Place all seed containers together under the same environmental factors.

(Poor drainage and overwatering might lead to rotted seeds and interesting but unwanted mold gardens.)

C. Procedures

1. Identify the manipulated variable; that is, decide which variable you want to systematically change. Then decide how you will go about changing it. Any one of the following might be the manipulated variable:

Kind of seed, or different brands of the same seed

Type of soil

Planting depth

Amount of water

"Pollutant," (such as salt, oil, detergent, plant food, pesticide) added in equal amounts to the water

Amount of "pollutant" added to water

Type of container

An environmental factor (such as light, temperature, humidity, noise)

2. Identify the responding variable, or what you think will happen. Will the number of seeds that germinate vary? The speed of germination? The height or appearance of the plants after a certain number of days? What results should be recorded?
3. Carefully analyze and identify the variables you wish to control. Also, decide on the means of control as described on page 188.
4. Reexamine your list of controlled variables and search for the less obvious variables. For example, is there any way of knowing the exact age or "health" of the seeds to be planted? Might the chemical content or temperature of the water vary from day to day? How constant are the environmental factors you are trying to control? Could intermittent sounds from an adjacent room have any effects on germination? Are there any supposedly uncontrolled variables that could be modified and then added to the list of controlled variables?
5. Decide on the extent of your search. How many samples should or can be tested? During how many days or weeks should or can data be collected? The answer to these questions will certainly influence the accuracy of your results and how much confidence you can have in the data. For example, suppose you saw the following advertisement in a magazine:

On the basis of test results, PATTON'S PAIN PILL relieved the pharyngitis of *all* sufferers sampled.* Buy today and swallow tomorrow with a smile.

*The sample population consisted of 5 persons tested December 1.

Table 7-6
SPROUT SEARCH VARIATIONS

Manipulated variable:
Controlled variables:
Responding variable:
Uncontrolled variables:

How much confidence would you have in Patton's pain pill? Perhaps you can recall some actual situations with similarly inconclusive test data. The greater the sample, and the more times the same conditions have been reproduced, the more reliable the results.

6. Decide, also, how to collect and organize the results. If possible, use charts, tables, or graphs.

7. After you have completed your search, share your summarized results with others. (See Table 7-6.) Compare and evaluate the variables which were investigated.

D. Extension   Usually the results of one investigation evoke questions for additional investigations. If this is true in your case, what other effects might be tested? Which are both feasible and worth testing? What additional responding variables can you investigate?

Select a population of manageable animals such as mealworms, ants, beetles, snails, crickets, isopods, guppies, or crayfish. Investigate responding variables related to food or habitat. What variables might be manipulated? What variables should be controlled?

Additional Considerations on Variables

"Did you control your variables today?" may never become a popular slogan, but it is one of the most important questions for anyone searching for explanations of causes and effects. The question could be asked in relation to any of the following possible events:

A strange new epidemic breaks out in a large city. The effect, or responding variable, is identical in all cases, but there may be thousands of possible causes.

The fish in a certain body of water are disappearing. What might be the cause or causes?

A chemical engineer discovers a material which can withstand incredibly great heat and pressure. Is its existence a unique accident? Can additional samples be made? What makes it so different from any other known material?

Changes—some expected and many unexpected—are occurring at all times in the natural world. What are the relationships among the changes? What factors influence other factors?

Perhaps the examples noted above seem far-fetched to you. Here are some examples that may be closer to home:

My cheese soufflés have always been a success. What happened this time?

It was much easier to mow the grass last time than this time. Why is it so much more difficult today?

This puppy is losing weight instead of gaining. What could be the cause?

Her car gets 27.6 miles per gallon of gasoline; mine gets only 14.3. How can I increase my mpg?

Last year's garden was a success. What went wrong this year?

My fresh vegetables always end up rotting. What will make them last longer?

Our roof was guaranteed for 20 years; it lasted all of 20 days. Why?

Although you may not be able to solve any of these problems completely, or even partially, you should now be able to approach them more rationally. You can attempt to examine and identify the variables involved. The more you are able to control all but two variables—the one you manipulate and the one you subsequently observe changing—the better your chances are of determining how one variable effects the outcome.

The skill of controlling variables is not easy to master. Preoperational learners and most concrete-operational learners find it impossible or extremely difficult to identify and separate the different types of variables—those that one changes (the manipulated variables), what happens as a result of the change (the responding variable), those that are kept the same in all cases (the controlled variables), and those factors that stem from human error or mechanical flaws (the uncontrolled variables). Most people at all ages or stages of intellectual development have the tendency—or are at least tempted—to try to do more than one thing at a time. That is, they often fail to isolate and vary only *one* factor systematically. Do not be surprised if the initial attempts at controlling variables are rather primitive and haphazard. Such stumbling about is necessary. But to *teach* the terms *manipulated variable*, *responding variable*, and *controlled variables* to children artificially— that is, to *show* them—is an exercise in futility. *Only as learners become dissatisfied with their own chosen investigatory procedures, and experience disequilibrium, should the technique of controlling variables be suggested.* You can facilitate children's acquisition of this skill more by purposefully directing questions than by dictating procedures. The preceding activities should not only have made you more familiar with types of variables and techniques for controlling them, but also have given you some ideas for directing questions in the search for clearer explanations.

Recording and Interpreting Data

Data processing is such a common action that it might seem strange to include it among the skills to be considered. In fact, this might also be true of several of the other skills—such as observing, communicating, and predicting. Why bother?—Because it is helpful and often necessary to reexamine the obvious. There is always a danger that familiar concepts might lose their potency. The opportunity for retrospection can often deepen not only your understanding of a specific concept but also your use of a skill.

When and how do you deliberately use the combined skill of recording and interpreting data? Think about the facts and figures that need to be interpreted in any of the following instances:

Examining the bargains advertised in a newspaper

Listening to the news, or to a discussion.

Anticipating the arrival of friends for the weekend (with their four young children and two dogs)

Selecting a new home, car, doctor, or career

Preparing a shopping list

Planning a building project

Completing the nine major activities presented in Chapter 6

Activities

Since you've already had considerable experiences with the skill of recording and interpreting data, only two activities will be presented in this section. As you complete these activities, try to note how the skill of recording and interpreting data is integrated with all the other skills presented so far.

Ready . . . Aim . . . Fire!

A. Introduction As a child, how many times were you cautioned, scolded, or even punished for shooting rubber bands? This activity gives you the chance to shoot rubber bands with impunity. It also will provide an outlet for latent talents, release energy, and allow you to integrate many science processes—of which recording and interpreting data is only one. The general idea is derived from Charles E. Armstrong's article "Shoot-Out at Claypit Hill."* The activity can be completed individually; however, a three-person team is recommended.

B. Materials (1) A wooden "shooter," as shown in Figure 7-2a or b; (2) a collection of rubber bands of various sizes; (3) a meter stick or tape; (4) masking tape; and (5) a piece of aluminum foil approximately 30 cm square.

Decide which shooter you can construct or assemble.

If you choose Figure 7-2a, carefully analyze the version of Armstrong's shooter shown there. The approximate dimensions for the three main pieces are labeled: a platform, the "barrel," and the angle support. Five or six holes approximately 1 cm apart should be drilled vertically through the front end of the barrel. (Position the first hole about 4 cm from the end.) Hammer a nail about 2 cm long with a very small head into the front end of the barrel, centered approximately 4 mm from the top surface and extending outward about 3 mm.

*Reproduced with permission from *Science and Children*, April 1971, p. 30. Copyright 1971 by the National Science Teachers Association, 1742 Connecticut Avenue N.W., Washington, D.C. 20009.)

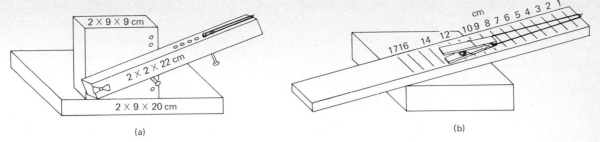

Figure 7-2 Two types of rubber-band shooters.

Another five or six holes approximately 1 cm apart should be drilled laterally through the front end of the angle support. The final hole should be drilled laterally through *both* the end of the barrel *and* the rear corner of the angle support. Attach the angle support to the edge of the platform, as shown, by hammering in at least two sturdy nails from the underside of the platform. Use a long bolt with a wing nut to attach the rear end of the barrel to the rear corner of the angle support. The bolt will serve as a hinge for the movable barrel. A support for the barrel can be made by removing the pointed end of a long, strong nail or by using a thin wooden peg. Position the rod through one of the holes in the angle holder so that it extends enough to support the base of the barrel. (If the barrel tends to move, tighten the wing nut.) Use another, shorter, strong nail (with the point removed) or a wooden peg as a trigger. The trigger should be placed through the bottom side of any vertical hole in the barrel and extend about 2 mm on the top side or simply enough to hold the rubber band in position. Connect a rubber band to the small nail protruding from the front end of the barrel and the extended trigger. Use a sudden downward pull to release the rubber band, and observe the flight of the band.

If you decide to assemble the shooter shown in Figure 7-2b, use a board approximately 50 cm long, 7 cm wide, and 2 cm thick. (The dimensions for either shooter do not have to be exactly as described.) Hammer a nail about 2 cm long with a very small head into the middle of one end of the board, centered about 3 mm or 4 mm from the top surface. The nail head should extend outward about 3 mm. Mark off the centimeter measurements as illustrated. Place a book or any firm object under the front end of the board to establish and vary the desired angle. The back end of the board should be resting on the floor; it can be held securely with your knee. It can also be taped to a table surface. A protractor could be used to measure the various angles selected. The trigger for this shooter is a clothespin with the strongest spring you can find.

Although the shooter pictured in Figure 7-2a is more difficult to build, it provides for better control of variables, especially of the angle positions and the release positions. The other shooter assemblage provides for a much wider variety of positions; but there is a greater need to be attentive to controlling the variables. (Decisions, decisions.)

C. Procedures

1. Select an open area to be used as your shooting range. The area should be

at least 6 m long and 1 m wide and be located in front of a closed door or bare wall. (If you use a closed door, it would be wise to post a warning sign on the opposite side, so that no one will walk into a rubber-band firing squad.) Decide if you want to operate the shooter from the floor or from a table.

2. Always load and aim the shooter toward the door or wall. Take any number of practice shots in order to become more familiar with the effects of using various angles, release points, and sizes of rubber bands.

3. Discuss the variables involved. What is the responding variable? What variables can be manipulated? What variables might be controlled? What uncontrolled variables might be present?

4. Examine this challenge:

> The 30-cm-square piece of aluminum foil will be taped by a nonparticipant at any position on a door or on a wall area comparable in size to a door. The target-positioner will also designate how far from the target the shooter should be placed: this distance can be anywhere from 1 m to 6 m. *Your task is to fire a rubber band onto the target in three tries or fewer.*

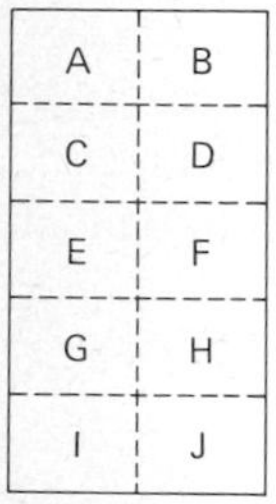

A	B
C	D
E	F
G	H
I	J

Figure 7-3 Locating rubber-band hits.

5. What strategy can you develop to meet this challenge? What system for organizing data might be used? Figure 7-3 shows one way of recording the locations of the rubber-band hits on a door. Table 7-7 shows one possible way of organizing data; use it unless you can devise a more suitable system.

6. Collect data on fifteen to thirty different shots, or enough representative data from a variety of conditions to give you confidence to meet the challenge. In order to try to control more of the variables, one person should act as the shooter, another as the measurer, and the third as the recorder of data.

7. When you have reached the necessary level of self-assurance, ask someone who has not participated in the activity to (a) tape the aluminum target to the door (or a wall area equivalent in size to a door) and (b) specify the distance of the shooter from the target (not to exceed 6 m).

8. Remember—you have only three tries in which to hit the target; therefore, interpret your data carefully. Then evaluate your performance and share your results with other rubber-band shooters. (If interest in the Crawling

Table 7-7
RECORD OF RUBBER BAND TRIAL SHOTS

DESCRIPTION OF RUBBER BAND USED	VARIABLES CONTROLLED	DISTANCE OF SHOOTER FROM TARGET	RELEASE POSITION	ANGLE POSITION	LOCATION OF HIT

Spool Marathon ever subsides, you might begin plans for organizing a Rubber Band Shoot.)

D. Extensions If a larger floor space is available, this activity could be modified to allow for longer projections of the rubber bands. The target could be laid on the floor anywhere within a range of 1 to 10 m. You might also try inventing (or encouraging your students to invent) a different type of rubber-band shooter.

If possible, examine the Teachers' Guide, the Student Manual, and the equipment for the fifth-grade unit "Energy Sources," developed by the Science Curriculum Improvement Study (SCIS). One section in the Student Manual for this unit deals with a fascinating multipurpose piece of equipment that can be made into a stopper popper—that is, a shooter of rubber or cork stoppers (1971, pp. 18–20). The Rand McNally SCIIS unit employs a similar apparatus—a catapult system (Thier et al., 1978). Although the skills of controlling variables and interpreting data are stressed in these units, the major emphasis is on the recognition of motion as evidence of transfer of energy. The units have an interesting assortment of materials (such as rolling spheres, paper airplanes, rotoplanes, ramps, and sliders) with many activities that involve interpreting data.

If you wish to branch off into some of the most basic principles of rocketry, trajectories, or actions and reactions, examine Learning Activity Unit 8, "Blast Off," in the module "Oceans and Space" of *Self-Paced Investigations for Elementary Science* (Katagiri, Trojcak, and Brown, 1976, pp. 83–95). Three investigations are presented on what makes a rocket or a balloon go, how thrust effects the distance that a rocket-balloon travels, and what influences the path of a projectile.

A few suggestions and comments might be helpful concerning the interpretation of data on moving objects in general. Some teachers or administrators turn pale (or red) at the thought of allowing children to propel any object in a classroom or an adjacent area, feeling that the confusion and the danger of injury are not worth the risk and that the efforts needed to control the children rather than the variables are too great. Such generalizations are not necessarily true; they are often based on inferences rather than observations. Once several of my college students and I directed an after-school science program for inner-city children in grades 1 through 8. Daily enrollment varied according to the menu of snacks supplied by a city catering service. We also offered a smorgasbord of activities somewhat similar to those you've already experienced. A modified version of the rubber-band activity was presented; and it was obvious that this was the best received and most valuable learning experience of all. The children finally saw the need to observe, measure, detect spatial relationships, predict, consider variables, hypothesize, and—most important—*record* and *interpret* their data. Their desire to solve the problem was far greater than the temptation to shoot at each other. The success of this activity did not suddenly convert these children into model students or budding scientists. But it did show us that children can have more self-discipline and self-determination if we gave them meaningful opportunities for responsibility along with reasonable guidelines.

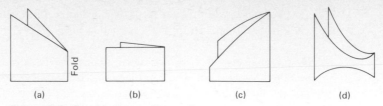

Fold

(a)   (b)   (c)   (d)

Figure 7-4   Possible "paper dragons."

Slide Systems Analyst Needed

A. Introduction   Many occupations require the skill of interpreting data. Teachers must exercise this skill many times each day as they plan and present the next learning encounters after evaluating the effectiveness of the preceding ones. Doctors must record and interpret data on bodily functions. People involved in the manufacture and sale of goods are constantly examining reports on input and output. There are, of course, countless other examples of the need to interpret data on the job. Sometimes, one must demonstrate the skill of interpreting data in order to get a job.

Companies and government agencies often announce the availability of contracts to individuals or firms which can meet certain specifications for a product or performance. This activity simulates that situation. It is related to "The Paper Dragon" in *Modular Activities Program in Science* (Berger et al., 1974, pp. 237-245).*

B. Materials   (1) A ruler-ramp marble roller, as shown in Figure 5-21, consisting of a grooved metric ruler, masking tape, an angle support, a marble, and a meter stick or tape, arranged on a *hard, smooth* surface such as a counter or table; (2) a protractor; (3) several 7.6- by 12.7-cm (or 3- by 5-in.) index cards; (4) scissors; and (5) a fairly small "bid box" containing 25 pieces of paper, labeled consecutively from 2 cm to 50 cm.

C. Procedures

1.   It is rumored that bids will soon be accepted for a "paper dragon" contract, and a skilled slide systems analyst—that is, someone who can determine how far a "paper dragon" will slide under certain conditions—is wanted. The exact design of the paper dragon has not been specified; the models shown in Figure 7-4 are some possibilities (they have been cut from small folded index cards).

2.   The paper dragon must slide a certain distance after being struck by one marble released from the ramp. The exact distance is not known either—but it must be somewhere between 2 cm and 50 cm from the attached edge of

*Adapted from C. F. Berger et al., *Modular Activities Program in Science*, teacher's edition. Copyright © 1974 by Houghton Mifflin Company. Used by permission of the publisher.

Table 7-8
PAPER DRAGON DATA

Variables controlled:				
MANUFACTURED VARIABLE	RESPONDING VARIABLE (SLIDE DISTANCE)			
MANIPULATED VARIABLE	TRIAL 1	TRIAL 2	TRIAL 3	AVERAGE

the ruler-ramp. The emphasis is on precision rather than power. An abundance of data must be collected if you are to be prepared for the contract bidding.

3. Your first task is to become familiar with the workings of the system. Make your own version of a paper dragon from a folded index card. Center the card directly in front of the bottom taped edge of the ruler-ramp so that the rolling marble is caught within the center fold of the paper and drags it along upon impact (hence the name *paper dragon*). Test a variety of combinations—vary the design of the paper dragon, the release position of the marble, or the angle of the ramp. (Remember that the more tries are made under the same conditions, the more reliable the results will be.)

4. Before you begin to systematically alter the conditions, identify what variables are involved. The responding variable will always be the distance the card slides. Therefore, you'll need to determine which variables are to be controlled and which variable is to be manipulated.

5. Decide what is the best way to collect and record your data. Table 7-8 is one possibility. Remember to alter only *one* variable at a time.

6. After you've collected and recorded your data, construct a graph (or several graphs) to illustrate the relationship between the variables. A graph will greatly aid in interpreting the results you observed. The distance in centimeters that the card slides—that is, the responding variable—should be recorded on the vertical axis. Data from the manipulated variable—such as the release position of the marble along a constant ramp angle or a constant release position from varied ramp angles—should be recorded on the horizontal axis. You might also graph the results of controlling the release point and the angle but varying the kind of paper dragon used.

7. Before you actually bid on the contract, examine the thoroughness of your data. For example, analyze the data shown in Figure 7-5 (page 198) from a competitor. Suppose the contract calls for a slide distance of 28 cm. What information does Figure 7-5 provide? What additional information is still needed? (A thorough data table might also include data on the angle of the ramp, the design of the paper dragon, the type of surface, the type of marble, and any other variables that were kept constant while only one factor was manipulated.)

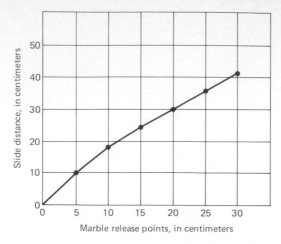

Figure 7-5 Paper-dragon data.

8. When you are ready to test your skill at interpreting data, reach into the "bid box" and take out one of the pieces of paper. This is your opportunity to try for the job of slide systems analyst. Demonstrate, and then evaluate, your ability to interpret data. If you can slide the card within several millimeters of the specified distance (in centimeters) at least seven times in a row, consider yourself a reliable slide systems analyst.

D. Extensions   Try testing a few more variables, such as (1) adding paper clips to the folded card, (2) using a steel sphere instead or a marble, or (3) using more than one marble to slide the card. The original specification could also be changed, to call for distance rather than accuracy.

If you are interested in relating the skill of interpreting data to actual environmental concerns, investigate the entire Teacher's Edition of the "Ecology" module of *Self-Paced Investigations for Elementary Science* (Katagiri, Trojcak, and Brown, 1976, pp. 1–96). Of the twenty-one major investigations presented in this module, twenty incorporate the skill of analyzing data. The topics include effects of insects on our food supply, problems of pollution, clean and unclean water, analyzing soil, and responses of certain animals to environmental factors.

"Mystery Powders" and "Rocks and Charts" are two excellent units in *Elementary Science Study* which place heavy emphasis on recording and analyzing data (1974). In "Mystery Powders," children record observations on the effects of water, heat, iodine, and vinegar on unknown white powders such as sugar, salt, baking soda, starch, and plaster of paris. Eventually, they are able to draw some conclusions by interpreting their data. The emphasis in "Rocks and Charts" is more on keeping records of observations than on learning the actual names of the twenty-one samples encountered. Children usually prefer performing activities rather than recording the resulting data. Organizing observations in a meaningful manner and then interpreting the data can become more significant when a challenge or a need to use the information is presented.

Additional Considerations on Data Processing

Interpreting data, like inferring, predicting, and hypothesizing, can aid in the search for supportive evidence and lessen the tendency to jump to conclusions. Like observing, distinguishing spatial and temporal relationships, classifying, and communicating, interpreting data enables one to make sense out of chaos, order out of disorder. This skill requires both analysis and synthesis—distinguishing the relevant from the irrelevant and forming a meaningful composite from which decisions, identifications, and supportable conclusions can be drawn.

Operationally Defining

Throughout much of your official schooling, you've probably been asked to learn a great many formal or dictionary-type definitions. Such definitions probably tended to be abstract—only slightly related to your actual experiences of the object or event defined—and to be easily forgotten. An operational definition is significantly different from a dictionary-type definition. Examine the following two examples:

Definition A; "Oxygen is an element composed of atoms having atomic number 8 and atomic weight 16."

Definition B: "Oxygen is a gas that causes a glowing splint to burst into flames [what you observe] when the splint is placed [what you do] in a container of the gas."*

The first definition sounds impressive and seems clear, but it communicates very little to those outside the field of chemistry. To the nonchemist, it may even squelch any further interest in the topic of oxygen. This type of definition often pervades science textbooks, lessons, and even courses. Formal definitions certainly have a meaningful place in learning. But if they occupy the initial position prematurely—that is, if they are presented *before* the learner has actually experienced their meaning—they can stunt further investigations and discoveries. They can violate the premise that experiential learning should precede language acquisition.

The second definition is far more meaningful to the learner; it is based on personal perceptions and actions—that is, on *interiorized* experiences. This is an example of an *operational definition*—a statement which describes an object or event by describing what operations or actions were performed and what was observed.

Usually, a wide variety of operational definitions are possible since one person's perceptions and actions in experiencing a concept may vary significantly from another's. What's more, one's own operational definitions might change and be refined over a period of time, depending on the type of experiences

*Permission to reprint from *Science—A Process Approach, Commentary for Teachers*, American Association for the Advancement of Science, p. 156, 1970.

encountered. For example, the operational definition of *teaching science* would be quite different for a sixth-grader, a college student enrolled in a course on teaching elementary science, a first-year teacher, and a veteran teacher. But if each individual was able to describe what actions were performed and what was observed in the teaching of science, communication and understanding would be greatly enhanced.

The skill of operationally defining often seems strange and foreign to adults initially. Somewhat paradoxically, children tend to use the skill quite naturally. Ask a child in the primary grades to define magnetism, and a likely response is: "Magnetism is when I put the magnet near some paper clips [what was performed], and the paper clips stick on [what was observed]." Such a response is frequently "corrected" by the teacher: "No; magnetism is a property or quality of magnets . . ." In Chapter 1, I quoted a rather extraordinary operational definition of scientists given by some first-grade children: "Scientists are people who make bombs [actions performed]; and those bombs kill people [observations or inferences]." These definitions of magnetism and scientists obviously reveal the children's limited understanding, but they also reveal personal, honest, and immediate viewpoints. Once this is recognized, a teacher is in a much better position to provide additional broadening experiences.

Although you may not have been aware of it, you've already been exposed to the need for operationally defining terms or events. In the activities on classifying you had to determine a standard for comparison or establish which objects had a certain characteristic and which didn't. If you responded to the encounter 4 on page 146 in Chapter 5, your answer was probably an operational definition of distinguishing spatial and temporal relationships, classifying, or communicating. In "Bouncing Ball Baffles" (page 161 in Chapter 6), you needed to determine the point at which the bounces were too rapid to count. In "Burned Out" (page 164 in Chapter 6), the burning time of the candle under the inverted jar also had to be operationally defined.

Activities

The following activities will provide you with some opportunities to consciously develop the skill of operationally defining.

Cohesive Communication

A. Introduction Have you ever encountered formal definitions from the physical sciences which seemed to have little meaning—*energy* as the capacity to do work, for example; *work* as the result of a push or a pull; *electricity* as a flow of electrons; *conduction* as the transfer of electricity, heat, or sound through matter; or *friction* as the resistance to movement? You may have found it far easier to memorize these relatively simple definitions than to understand them. Why? Some possible reasons are (1) that many of the descriptors were unclear or highly specific in meaning; (2) that you did not bring enough experience to those terms for them to have real meaning for you; (3) that, since you had not formulated the definitions yourself, they were really not "yours"; (4) that the definitions might have been presented prematurely, *before* the concepts were experienced (creating a situation similar to "learning" and quickly forgetting nonsense syllables). If such terms had been operationally defined, their meaning would have been much clearer.

You can see this for yourself in this activity on the concept of cohesion. The usual formal definition of *cohesion* is "that force by which the same kind of molecules are held together." By observing the results of the following procedures, you should be able to operationally define *cohesion* more clearly.

B. Materials   (1) A 30- to 40-cm strip of wax paper; (2) seven small baby-food jars or small glasses; (3) seven medicine droppers; (4) 20 ml of rubbing alcohol; (5) 20 ml of soapy water (add about 5 ml of liquid soap to 15 ml of water); (6) 20 ml of water; (7) 20 ml of liquid soap; (8) 20 ml of cooking oil; (9) about 10 ml to 20 ml of appliance or machine oil; (10) masking tape or any material for labeling each container; (11) a sheet of fine-lined graph paper; and (12) at least seven toothpicks. (If any of the suggested liquids are unavailable, substitute others, such as milk, coffee, liquid shampoo, mouthwash, or perfume.)

C. Procedures

1.   Pour the designated amounts of each liquid into separate containers; then label the contents and numerically code them from 1 to 7. In order to avoid contaminating the liquids, place a separate medicine dropper in each container of liquid.

2.   You'll also need a method for organizing your observations of samples of the different liquids on the wax paper. Figure 7-6 is one suggestion, but it can be modified as you see fit. The dots in the columns indicate where the

Figure 7-6   Examination form for drop data.

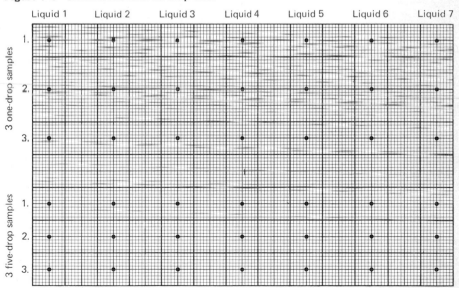

drops could be placed so that the surface areas covered by one-drop samples and by five-drop samples of each liquid could be compared more easily. The labeled graph paper should then be placed *under* the wax paper. If the wax paper tends to curl, tape the edges so that the entire surface is as flat as possible.

3. Before releasing one drop of each liquid onto the wax paper three times, think about the variables. Try to control the height of the dropper by holding it about 1 cm directly above the dot under the wax paper. Also, try to release each drop of liquid from the medicine dropper in a consistent manner.

4. Collect the data suggested in Figure 7-6. First complete the three one-drop observations of each liquid. How do the drops compare in terms of surface covered and height? (View the drops from different positions, especially from an eye-level horizontal position.) What hypothesis can you formulate at this point concerning the liquids' capacity to hold or cling together?

5. Observe the results of three separate five-drop samples of each liquid. How do these results compare with the outcomes of the one-drop samples? Reevaluate your previous hypothesis: do the new observations demonstrate the presence or the absence of supportive data?

6. Use a separate toothpick for each liquid drop sample and *very gently and slowly* move the second five-drop sample to the third five-drop sample. Carefully observe the results of each merger. What differences and similarities did you note among the results?

7. Now try to gently drag each entire third one-drop sample to the first five-drop sample. How do the results support your original hypothesis?

8. On the basis of what you've done and what you've observed, write your own operational definition of *cohesion*. Compare your definition with others' definitions and determine which is the clearest communication. Which operational definition would best describe cohesion to someone who had never attempted to define the concept?

9. Compare your operational definition with the following one, based on another experience: "Cohesion exists when a container is filled with water but more drops of water can be added [what was done] so that the surface of the water is eventually 2 mm higher than the top of the container [what was observed]." How is this operational definition similar to the one you formulated?

D. Extensions   Design some activities which could lead to operational definitions of evaporation, water vapor, condensation, or the most absorbent paper towel.

You can also try to formulate an operational definition of *adhesion*—that is, the force by which unlike molecules bind together when in contact. If you need some ideas, examine the suggestions from "Kitchen Physics" in *Elementary Science Study* (1974, p. 17 or pp. 71–80). For example, use a heavy piece of cardboard (approximately 25 cm by 40 cm) which can serve as a runway for the drops. Cover one side of the cardboard with wax paper and set it on several thicknesses of newspaper. Then place three-drop samples of each liquid previously tested in a line about 2 cm from one of the narrow edges of the cardboard. Raise the end with the line of drops to about a 45° angle and observe the race of the liquids to the opposite end. Order the results from fastest to slowest, that is, from least adhesive to most adhesive. Repeat these procedures with different surfaces such as aluminum foil, plastic wrap, and the uncovered cardboard. Include both the actions you performed and what you observed in order to formulate an operational definition of *adhesion*.

The Best

A. Introduction   As a consumer or as a traveler, how many times have you seen the title "best" bestowed upon products or places? "Bests" run the gamut from airline to zoo; in just about every category imaginable, there's an attempt to proclaim "the best." But often the title seems to be a token. It would be a fascinating experience to ask everyone who calls something "the best" to operationally define the meaning of *best*.

This activity is your opportunity to develop an operational definition for *best mealworm backer-upper*.

B. Materials   (1) Several thicknesses of newspaper to cover your work area; (2) a magnifying lens; (3) various liquids—those used in the previous activity or any others of your choice; (4) separate medicine droppers for each liquid; (5) various kinds of powders, such as salt, baking soda, pepper, and spices; (6) flattened toothpicks for each powder; (7) various samples of fruits and vegetables; (8) a small, clear drinking glass; (9) some matches; (10) a drinking straw; (11) a metric ruler; (12) ten to twenty healthy mealworms in a container. Mealworms can be obtained from most pet stores and some bait shops. They are the larvae or the worm-like stage of those common small black beetles which are often found underneath dead leaves, logs, or rocks. (Mealworms are harmless, and neither smelly nor slimy, though they may tickle a little.)

C. Procedures

1. Basically, this activity will be a study of various stimulus-response situations. But before you deliberately apply any stimulus, you should first observe the normal physical features and behavior of mealworms. (This is a sound practice to follow when studying any living organism.) Try to make at least ten to twenty observations of a mealworm without a magnifying lens and then with one. View the mealworm from different vantage points. If you have difficulty observing its underside, place it in a small glass and raise the glass above your eyes.

2. Put the mealworms aside; then test your observing skill by drawing a picture of one *from memory*. Compare your drawing with the mealworm shown in Figure 7-7a (page 204). Next, challenge your observational powers: observe just one mealworm so carefully that you could identify it in a crowd; then put it in with a group of ten to twenty others and identify it.

3. Observe the mealworm's normal behavior—how it walks, how the segments interact, where it travels, whether it seems to have any natural preferences,

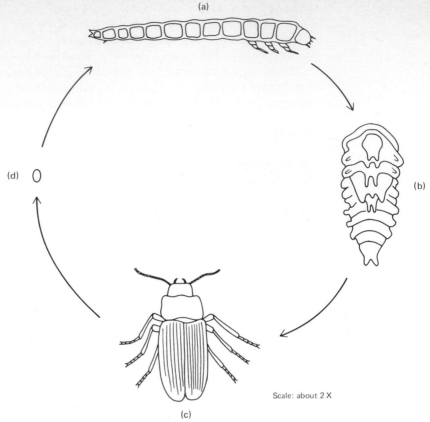

Figure 7-7   Stages in the life cycle of the mealworm: (a) larva (side view), 4 to 5 months; (b) pupa (underside view), 1 to 3 weeks; (c) adult (top view), 2 to 4 weeks; (d) egg (side view), 1 week.

how its antennae respond, how it reacts to obstacles, whether it travels at different speeds and in different directions. How would you describe the normal behavior of your mealworms? Discuss your descriptions with others in order to avoid the possibility of generalizing from some deviant behavioral patterns.

4.   Once you feel that you know both the physical features and the behavioral characteristics of your mealworms, you are ready to attempt the major task: To discover what will cause a mealworm to back up. Before you begin to organize your plans or formulate a testing strategy, note the caution below:

Be kind to your six-legged friend.

Nerve endings or chemical receptors cover the mealworm's entire body. Therefore, *never* place any testing substance directly on the mealworm.

Table 7-9
MEALWORM REVERSAL TESTS

Controlled variables:		
MANIPULATED VARIABLE (WHAT YOU DID)	RESPONDING VARIABLE (WHAT THE MEALWORM DID)	RANK ORDER

5. Examine the testing substances and materials suggested previously and add any of your own choices to the collection. Infer which items or methods might cause the mealworm to back up when placed in its path or proximity. Then test at least five to ten of the most likely substances or techniques many times. Use a format such as Table 7-9 to record your data. Again, identify the controlled variables. Take care that you don't overwork the mealworms or bombard them with too many stimuli. Give them time to recover or adjust from one encounter before beginning the next.
6. Interpret the data and rank-order the substances and methods from best reverser to poorest reverser. Then construct an operational definition of *best mealworm backer-upper*. How does your operational definition compare with the definitions of others?

D. Extensions Determine how long you can maintain a mealworm population. Keep your mealworms in a tall container with smooth sides, like a 1-qt jar or a coffee can. Add several handfuls of breakfast cereal—bran is especially good. Place a thin layer of shredded newspaper or paper towels on top of the cereal. Mealworms require very little moisture other than what's in the cereal. However, their growth and development can be slightly accelerated if a small piece of succulent fruit or vegetable, like an apple or a potato, is added once a week. When the cereal becomes powdery, add a fresh supply. If any mold growth appears, you've probably been adding too much moisture, and a new habitat should be prepared. Briefly observe your mealworm population every several days or at least once a week and record any changes that occur. If you're patient and persistent, you'll discover some subtle variations along with some remarkable transformations.

Figure 7-7 illustrates the complete life cycle of a mealworm. The pupal and adult stages are relatively easy to observe. However, it takes a bit of skill to locate the eggs of the next generation. These are only slightly over 1 mm in length. They are white oblongs or ovals which have a sticky covering when fresh. Therefore, they're easily covered by and lost amid the powdery remains of the cereal. With a hand magnifier and some perservance, you might eventually be able to locate larvae as small as 2 mm or 3 mm in length. The life cycle of the mealworm is an example of complete metamorphosis. (Butterflies also exhibit complete metamorphosis. Many insects undergo incomplete metamorphosis; for example, there is no pupal stage in the development of grasshoppers, dragonflies, and cockroaches. In termites, all metamorphosis occurs within the mother termite's body, so that newborn termites look like miniature adults.) After you've observed the life cycle of the mealworm, write an operational definition of *complete metamorphosis.* You may decide that it's easier to operationally define concepts related to physical science than concepts related to biology. Biological operational definitions tend to deal more with what you observe than with what operations you perform.

If you wish to conduct additional studies on mealworms, examine the unit "Behavior of Mealworms" in *Elementary Science Study* (1966). Although the emphasis is not on operationally defining, you can glean some additional ideas for activities which can lead to operationally defining the mealworms' best food, best habitat, and best accelerator (rather than best reverser). From the investigation "Climb Every Mountain" of the module "Plants and Animals" in *Self-Paced Investigations for Elementary Science,* an operational definition can be derived for the mealworms' best climbing surface (Katagiri, Trojcak, and Brown, 1976, pp. 22–25).

If you want additional practice, carry out the necessary procedures and collect the data for operationally defining other "bests"—such as best paint remover, best recipe for brownies, best plant food, best lubricant, and best means of conserving energy.

Going for the Gestalt

A. Introduction It is often difficult to pull ideas or procedures together into a meaningful gestalt or see how parts fall into an understandable pattern. Overall reasoning and methods are usually unique for each person. Sometimes suggestions and directions from others help; sometimes they do not. Usually you'll never know until you try them.

I have a theory that may help you discover a gestalt for the process skills you've been using. The theory is simple: *The interiorized learning of the process skills can be demonstrated in the form of operational definitions.* Certainly, there are other ways of demonstrating that learning has occurred or that the skills have resulted in real adaptation to encounters. The ability to distinguish examples from nonexamples is one proof of learning; so is the ability to apply what is learned to new situations, especially situations in which previous learning must be used to solve new problems. But sometimes these criteria are time-consuming or not easily manageable. Operationally defining is a much more economical criterion for learning. Even though operational definitions are not uniform and are subject to change, they do represent personal perceptions and actions. Because they are generated by the learner rather than taken from a book, they are more honest and accurate indications of actual understanding.

In this activity you can use operational definitions to assess your own learning.

B. Materials The following sections on the process skills:

Observing, Chapter 4, pages 81 to 88
Distinguishing Spatial and Temporal Relationships, Chapter 5, pages 114 to 122

C. Procedures

1. Draw upon either specific activities or the general characteristics of the examples to construct an operational definition of each of the nine skills.
2. When you're satisfied with your efforts, examine your products, share them with others, and decide if any modifications are needed. For example, how do the following operational definitions compare with your products?

Observing is using the five senses (if safe)—tasting, smelling, hearing, feeling, and seeing—along with measuring (what you do) to identify such qualities as sweetness, bitterness, sourness; mustiness, fruitiness; rancidity; pitch, volume; smoothness, roughness, bumpiness, prickliness; symmetry; length; mass (what you observe).

A spatial and temporal relationship exists when you release a drop of green food coloring into a glass of warm water and observe the color descend in long strings after 5 seconds, a larger concentration sink to the bottom in 15 seconds, the strings become more spread out in 30 seconds, and the water turn almost completely green after 2 minutes.

Communicating takes place when a person describes in writing a peanut's characteristics—such as: 3.7 cm long, with a 6-mm crack at one end, gray spots at the other end, and a 3-mm string-like projection—and then places the peanut with other peanuts not described and gives the written description to another person, and the other person is able to select the peanut that was described.

Hypothesizing is saying that a heavier bob will have a slower frequency than a lighter bob. After recording the frequency of the lighter bob for 60 seconds, no significant difference is apparent.

3. Ponder (or discuss) the pros and cons of my theory. Is the ability to operationally define a good indicator of real learning? A moderate indicator? A poor indicator? No indicator at all?
4. Consider these questions:
   a. In what ways does this skill encourage or discourage false accommodation?
   b. How did the construction of an operational definition for each skill enable you to be selective, to distinguish the relevant from the irrelevant aspects of the activities related to the skills?
   c. How might your perceptions of science be different if you had been expected to formulate operational definitions rather than memorize formal definitions?

d. When might the ability to operationally define the processes (or to operationally define in general) be most useful?

D. Extension   In *Why Am I Afraid to Tell You Who I Am?* John Powell states: "It is certain that a relationship will be only as good as its communication" (1969, p. 43). How is this statement related to the skill of operationally defining? How might this skill influence your communication and relationship with others—with your students, your colleagues, your friends, and the countless nameless people you encounter?

Additional Considerations on Operational Definitions

The skill of observing has been stressed consistently throughout Part Two as the beginning focus for doing science. The skill of operationally defining is not necessarily the terminal point, but it is a means of converging previous learning. As you select science encounters for children, it is important to be aware of the actions you hope to see them perform—that is, what is to be done and what should be observed as they respond to the learning encounters by observing, measuring, distinguishing spatial and temporal relationships, classifying, communicating, inferring, predicting, hypothesizing, controlling variables, recording and interpreting data, and operationally defining. As you deal with the daily encounters of living, all the process skills can also enable you to make more sense out of your world. In the final analysis, these processes can help you to better operationalize learning, rationality, and caring—some of the essential goals of science, for children and for everyone.

Encounters

1. Which of your daily activities require the use of controlling variables, interpreting data, and operationally defining? When should you have used these skills but failed to do so?
2. Analyze your buying habits. In the past, have you ever purchased a product mainly because it was "scientifically tested"? If you were to hear that said about a product today, what questions might you ask? How might you test the validity of the claims made about the product's performance or effectiveness?
3. In what ways do you agree or disagree with the inference that formal definitions of scientific terms can squelch further interest or stunt further investigations when they are given prematurely?
4. Examine some elementary science textbooks or curricular materials in terms of how the skills of controlling variables, interpreting data, and operationally defining are presented and implemented. If possible, observe the teaching and learning of these skills in actual elementary classrooms. Develop your own activities for these skills and try them out with children. Share the results with your colleagues.

5. What is your opinion of the appropriateness of the three skills presented in this chapter for various ages and grade levels? What are the bases for your opinion? (Refer to encounter 4 above and to Piaget's findings in order to substantiate your answers.)
6. Review all the process skills presented in Part Two. How are they applicable to classroom encounters in other subjects? How can they be applied to improve your own daily encounters?
7. Your choice.

References

Armstrong, Charles E.: "Shoot-Out at Claypit Hill, or The Rubber Band Unit," *Science and Children*, April 1971.

Berger, Carl F., Glenn D. Berkeheimer, L. E. Lewis, Harold T. Neuberger, and Elizabeth A. Wood: "The Paper Dragon," Modular Activities Program in Science, grade 5 teacher's edition, Houghton Mifflin, Boston, 1974.

Elementary Science Study, "Kitchen Physics," "Behavior of Mealworms," "Mystery Powders," and "Rocks and Charts," Webster Division, McGraw-Hill, St. Louis, Mo., 1974.

Katagiri, George, Doris Trojcak, and Douglas Brown: "Ecology," "Oceans and Space," and "Plants and Animals," *Self-Paced Investigations for Elementary Science*, Silver Burdett, Morristown, N. J., 1976.

Powell, John: *Why Am I Afraid to Tell You Who I Am?* Peacock Books, Argus Communications, Chicago, Ill., 1969.

Science—A Process Approach II, Module 71, "Seeds and Soap," 1975, and *Commentary for Teachers*, 1970, AAAS/Xerox Corporation, Ginn, Lexington, Mass.

Science Curriculum Improvement Study, "Energy Sources," Rand McNally, Chicago, Ill., 1971.

Thier, Herbert D., Robert Karplus, Chester A. Lawson, Robert Knott, and Marshall Montgomery: *Energy Sources,* Rand McNally, Chicago, Ill., 1978.

Unified Science and Mathematics for Elementary Schools, "Consumer Research—Product Testing," Education Development Center, Newton, Mass., 1976.

Suggested Readings

All the sources previously listed in Chapters 5 and 6 also contain ideas or implications related to the skills of controlling variables, interpreting data, and operationally defining.

The following modules for the intermediate grades from *Science—A Process Approach II* (Ginn, Lexington, Mass., 1975) can provide additional ideas for activities specifically related to the three skills presented in this chapter.

Module 62, "Climbing Liquids" (controlling variables)
Module 63, "Maze Behavior" (interpreting data)
Module 64, "Cells, Lamps, Switches" (defining operationally)
Module 65, "Minerals in Rocks" (interpreting data)
Module 67, "Identifying Materials" (interpreting data)
Module 68, "Field of Vision" (interpreting data)
Module 72, "Heart Rate" (controlling variables)
Module 74, "Biotic Communities" (defining operationally)
Module 76, "Limited Earth" (interpreting data)
Module 77, "Chemical Reactions" (controlling variables)
Module 81, "Analysis of Mixtures" (defining operationally)
Module 82, "Force and Acceleration" (controlling variables)
Module 88, "Environmental Protection" (defining operationally)
Module 89, "Plant Parts" (defining operationally)
Module 91, "Flowers" (defining operationally)
Module 93, "Temperature and Heat" (defining operationally)
Module 94, "Small Water Animals" (controlling variables)

The following units from *Elementary Science Study* (McGraw-Hill, St. Louis, Mo., 1974) which have not been previously mentioned incorporate the skills of data collection and organization as well as the identification and control of variables. Although the skill of operationally defining is not explicitly identified in these units it is present implicitly:

"Animal Activity"
"Brine Shrimp"
"Budding Twigs"
"Changes"
"Eggs and Tadpoles"
"Gases and Airs"
"Pond Water"
"Small Things"
"Starting from Seeds"

The following units from *Science Curriculum Improvement Study* (Rand McNally SCIIS, Chicago, Ill., 1978) also provide additional use of the skills discussed in this chapter:

"Relative Position and Motion"
"Energy Sources"
"Scientific Theories"
"Environments"
"Communities"
"Ecosystems"

PART THREE

PART THREE

OUR

initial foundation has been established. Science, learning, and children are dynamic enterprises. Each child is unique, yet children generally have certain characteristics as they progress through the stages of development. Their learning evolves from responding to encounters by using the thinking or process skills described in Part Two. Your learning will also be enhanced by these mental operations.

Although the process skills have typically been attributed to the types of actions scientists perform, by now I hope you've begun to see their wider applicability. Anyone—a child in grade school, a teenager in high school, a college student, a preservice teacher, a parent—can use many of these thinking skills to make more sense out of learning encounters. But teachers especially can incorporate the mental operations with their instructional skills. If the primary goal of education is to improve one's ability to reason, these thinking skills apply to all areas of learning. Therefore, they certainly apply to those actions of a teacher which are intended to facilitate learning—namely, instructional skills. In Part Three, the process skills will be related to important instructional skills necessary for science with children. The foundation will be strengthened; but the final decision to build upon it will be yours.

SCIENCE WITH CHILDREN
APPLYING INSTRUCTIONAL SKILLS

CHAPTER 8
OBSERVING AND INTERPRETING CLASSROOM INTERACTIONS

215

The process skills presented in Part Two are first and foremost *thinking* skills. They are applicable not only to the study of science but also to the entire educational enterprise. These thinking operations can become a means for adapting more rationally to the encounters of daily living. They can also be incorporated into the repertoire of skills a teacher needs in the classroom—namely, the broad range of instructional skills.

Relating Science Skills to Instructional Skills

There are numerous ways teachers can use the skills of science in the classroom. The following are only a few examples:

Observing the patterns and progress of learners, the latest curricular materials, signs of change, sounds over the intercom, odors from the lunchroom, the amount of perspiration after recess or gym.

Measuring students' achievements, amounts of material "to be covered" from the curriculum guides, the year's paper supply, the things to be manipulated in an activity

Distinguishing spatial and temporal relationships: determining what topics and materials can be scheduled within the limits of space and time, determining what items can be stored where and for how long, scheduling the time and places for group learning.

Classifying or grouping learners according to ability, interest, or behavioral patterns; organizing materials and supplies into a manageable system; sorting relevant from irrelevant suggestions in textbooks.

Communicating intentions, expectations, directions, encouragement, corrections, and satisfaction to students, parents, colleagues, administrators, and support staff; writing reports, asking questions; responding to others' questions.

Inferring why a usually vivacious student returned to the room so sullen, why faculty members became silent when a certain teacher walked into the lounge, what caused the mealworms to die, whether learning really occurred.

Predicting that today's lesson plan will finally end the confusion between two concepts, that real specimens will have more meaning than pictures in a book, that new concepts should not be introduced immediately before a holiday.

Hypothesizing about the best sequence for developing a particular topic, about whether more practice will always produce better results, about whether fewer questions are more effective than many questions.

Analyzing variables—time, distractions, fatigue, bias, favoritism, last year's labels; attempting to determine the reasons for success or failure.

Interpreting data on performances, on preferences regarding the next procedures, on the basis for grading, on the decision to promote or retain a student.

Operationally defining "success," "growth," "improvement," "failure," "learning" in general, or "effective" and "ineffective" instructional strategies.

The Importance of Observing Classroom Interactions

No one has ever identified all the skills an effective educator or scientist must be able to use; but clearly instructors and scientists have some skills in common. The importance of the skill of observing, in particular, has been stressed frequently as the starting point of science; and it is also basic when teaching and learning are viewed as an interactive interpersonal relationship. Before any instruction can begin, a teacher must gather observations on many questions: Who are these students? Where are they on the academic ladder? How might they respond to specific encounters? What skills and knowledge do they already have? What do they lack? What experiences do they need? What materials are available? Teachers who begin instructing before considering such questions often reflect a desire to teach subject matter rather than educate children. The preliminary observations a teacher collects about the students will provide a basis for the instructional decisions to be made.

It's interesting to note that people seldom refer to being "born into" any profession other than teaching and acting. You rarely hear the expression "born biologist," or "born engineer," or "born plumber." We feel that these professionals *learn* their trades—by reading, listening to lectures, and watching demonstrations, and also by direct experience with specimens, clients, or objects. But the same is true of teachers (and of actors too). Very few people can walk into a classroom (having walked out of one as a student several or many years ago) and immediately create a meaningful learning environment. It's not impossible, but it is improbable. Even an experienced substitute teacher will find persistent problems. Most people need some time to acquire a "feel" for the classroom by simply observing the components present as well as their interactions. The greater the variety and frequency of classroom observations, the better the inferences or predictions and, eventually, the more accurate the decisions. After spending time in classrooms you will be in a much better position to discover the many roles of a teacher—planner, presenter, facilitator, counselor, referee, director, judge, dispenser, comforter, and—the most important and all-encompassing role—decision maker. But in order to be an effective decision maker, you must first be aware of what's going on. More precisely, you must have a basis for your decisions. Then your subsequent actions can be performed and controlled more consciously. By learning to observe the components of the classroom and the behaviors of others, you can begin to conceptualize your preferred instructional style. You can also formulate more realistic expectations about children's learning. But these things are not accomplished automatically. If you observe in a classroom as a tourist on a tight schedule, little will be gained. The same is true if you observe as a penniless window-shopper. You'll need a reason or purpose for observing classroom interactions. Purposeless observing has little value; it tends to degenerate into boredom or isolated bits of trivia.

General Classroom Interactions to Observe

What aspects of classroom interactions are you most interested in observing? What is most worth observing in order to improve your science with children?

None of the activities you experienced in the previous chapters can compare with the complexity and intricacy involved in observing a real classroom of children encountering science. Perhaps the closest analogy was the activity "A Bug's-Eye View" (page 85)—but even then the number of objects and events were minimized. The strategy utilized in "A Bug's-Eye View"—that is, going from the general to specifics—might be useful for classroom observing. Usually when a land developer is investigating a possible building site, the entire terrain is studied. When a patient visits a doctor for the first time, a general examination is conducted. So, too, when you observe a class of children in action, it would be wise to view the total scene initially. Then you might try concentrating on any of the major components, such as the following:

People present—teacher or teachers, students, aides, other support staff, visitors

Encounters—what's being presented by the leader, or by the printed word, or by tape or film

Materials related to the encounters, or materials in competition, or simply materials present

Space in which the people and materials are situated

Time factors—time of day, or the impact of the schedule to be followed, or the pacing of the procedures or directions

Interactions among the factors which result in learning, reinforcement, dead ends, chaos, or further challenges

These are just a few of the general elements present in every classroom; thousands of specifics are possible. The difficulty of observing the seemingly countless components, as well as the interactions among them, can be alleviated only by determining answers to a twofold question: *What and why do you wish to observe? How* you observe a classroom depends on the answers to this question.

In what type of classroom would you like to observe science with children? Self-contained? Open? A classroom with several age levels? A classroom with team teaching? Besides the general components listed above any of the following also could be considered:

Verbal behavior
Nonverbal behavior
Types of questions
Types of responses
Disciplinary techniques
Modes of instruction
Levels of thinking
Use of the process skills
Affective behavior
Techniques of motivation
Use of time
Use of materials

Are you interested in observing just the teacher's behavior, just one student's behavior, the interactions within a small group, or the total class? Once you decide on *what* and *why* you wish to observe, you have several choices of technique. These are discussed in the following section.

How to Observe More Systematically

Effective observing requires a decision on how to observe—that is, the choice of an observational instrument or instruments.

You can analyze some observational instruments developed by others and select one or a combination of several which best suit your purpose or purposes. Or you can develop your own instrument and methods for classroom observing. Or you can select the best of both worlds—that is, incorporate others' ideas with your own instrument.

One of the most thorough sources for information on observational instruments developed by others is *Mirrors for Behavior III: An Anthology of Observation Instruments* (Simon and Boyer, 1974). This is a collection of ninety-nine different observing instruments. Descriptions of the categories for each system are presented, along with the setting and the subjects for which the instrument was intended, procedures for collecting and coding data, and an explanation on the overall use of each system. Four systems from *Mirrors for Behavior III* that are especially worth considering are Ned A. Flanders's "Categories for the Flanders System of Interaction Analysis "(FSIA); Elizabeth Hunter's "Categories for the Revised Verbal Interaction Category System—Science" (revised VICS); Bob B. Brown's, "Categories for the Teacher Practices Observation Record" (TPOR); and Charles C. Matthews's "Categories for the Science Curriculum Assessment System—Student" (SCAS). Some of the major characteristics of these four systems are shown in Table 8-1. Each system is basically a shorthand technique for recording events of particular interest. Observations can be recorded in the form of tallies to indicate the frequency of

Table 8-1
COMPARISON OF FOUR OBSERVATION SYSTEMS

OBSERVATION INSTRUMENTS	PEOPLE OBSERVED			MAJOR CATEGORIES				
	TEACHER AND STUDENTS	TEACHER(S) ONLY	STUDENT(S) ONLY	AFFECTIVE	COGNITIVE	ACTIVITIES	MATERIALS	SOCIAL STRUCTURE
FSIA	*			*				
REVISED VICS	*			*	*		*	
TPOR		*			*	*		
SCAS			*			*		*

occurrences. Or a more systematic approach can be used—recording each identified event every 3 or 5 seconds, or each time the category changes within less than 3 seconds. Longer time segments can also be used.

Generally, if instruments developed by others are selected, a considerable amount of practice is required to use them effectively. The more clearly recognizable the categories, the easier the recording. The descriptors for each category are or should be operational statements of the factors being considered. But even with the clearest of operational definitions, it is important to realize that each person perceives differently, and that what is recorded should be *observations,* not inferences. The records are descriptive rather than prescriptive—that is, the records should represent what is (or what was) according to your perceptions, not what ought (or ought not) to be or to have been. These points will become more meaningful if you apply them to an instrument of your own design.

In considering how your observations might best be recorded, classified, interpreted, and communicated, you have several important concerns: How can you best attend to your original purpose or purposes without becoming distracted by competing variables? How can the observations be recorded briefly and accurately so that you are not overwhelmed with bookkeeping chores? How can objectivity be maintained? These certainly are not easy questions to answer. Probably the best anyone can do is to make honest attempts; definitive answers are too much to hope for. There is no single, foolproof instrument, but there are some general techniques worth using. Most observation instruments fit into one of three categories: descriptive records of samples, checklists, and coding systems. Some examples will be presented for each type.

Descriptive-Record Technique

This entails an open-ended listing of brief descriptions of the observations. The technique is very similar to that suggested in Table 7-9 (page 205) for recording observations of mealworm tests. For example, you might observe classroom

Table 8-2
DESCRIPTIVE RECORD OF TEACHER-STUDENT INTERACTIONS

WHAT THE TEACHER DID	HOW THE STUDENTS RESPONDED

interactions in terms of what the teacher does and how the students respond. One possible form for recording the data is shown in Table 8-2. Brief descriptions of classroom examples would be recorded under the appropriate category. The chart could easily be modified to include additional information, such as the length of time spent on certain actions or the frequency of occurrences. After the data are recorded, they could be analyzed in various ways. For example, what percentage of time is related to certain activities? What is the rank order of the five most frequent behaviors by students? What behavior of the teacher elicited the most responses from the students? Which actions by the teacher resulted in positive behavior by the students? In negative behavior? In neutral behavior?

There are countless ways to construct a descriptive record of observations. Another format is shown in Table 8-3. In using this format, try to be selective in your choice of what is to be observed. Choose a rather general topic with wide applicability that you consider worth observing and possible to observe. If your choice is too specific or limited, the possibility that it will actually occur may be remote. For example, more opportunities would be available to observe methods of motivation, use of materials, or use of the process skills than to observe the introduction of a specific concept such as magnetism or the use of a certain piece of equipment such as a microscope. Your record of descriptions, however, should consist of *specific* examples of the more general topic you selected.

Table 8-3
OPEN-ENDED DATA COLLECTION FORM

Observer: _____ Date: _____ Time: _____ to ___

Subject(s) (or components) observed: _____

Activity or lesson: _____

What is to be observed: _____
_____
_____

Evidence of related behaviors   Frequency
_____   _____
_____   _____

Table 8-4
OBSERVATONS OF PIAGET'S INTERACTION MODEL

EVENTS	EVIDENCE (WHAT WAS DONE)	RESULTS (WHAT WAS OBSERVED)
ENCOUNTER		
FITTING IN; ASSIMILATION		
REINFORCEMENT		
DISEQUILIBRIUM		
"DEAD END"		
CHANGE; ACCOMMODATION		
ADAPTATION		

Another variation of the descriptive-record technique is shown in Table 8-4. Observations could be made of a child's reactions during a specific science learning encounter in terms of Piaget's interaction model. (If necessary, review Figure 2-1, page 26, and the explanations of the model in the text.) The observations recorded will result in your own unique operational definitions of the particular components of the interaction model. This type of observing-recording experience could be even more interesting if several observers recorded different individual interactions with the same science lesson and then compared their notes.

The elements in a descriptive record are rather easy to construct because of their general nature. Little time is required to prepare this sort of instrument, but considerable time is needed for recording the observations during the actual classroom interactions. The opposite is true for the next observing technique, the checklist.

Checklist Technique

The checklist technique is useful if you can identify in advance specific examples of a general category you wish to observe. Recording observations then simply entails tallying the occurrence of each specific instance. For example, suppose you are interested in observing the different modes of instruction used during a science learning encounter. What are the various possibilities? These must be clearly identified and arranged in such a way that a check mark or tally can be used to indicate their occurrence.

Table 8-5 shows one possible arrangement. This format could easily be modified to include observations of several different science classes by simply adding more horizontal divisions. A comparison could then be made of the frequency of modes of presentation in a variety of settings or grade levels. Time intervals could also be added, as shown in Table 8-6, which deals with observing the uses of materials.

Vary the examples of uses of materials (or modes of instruction) according to your special interests. These examples are intended to serve only as

Table 8-5
MODES OF INSTRUCTION

	LECTURE	DISCUSSION	DEMONSTRATION	READING	GROUP ACTIVITY	LEARNING CENTERS	GAMES	LAB SHEETS	ORAL REPORTS	FIELD TRIPS	TEAM TEACHING	GUEST LECTURER	ETC.
FREQUENCY													

Table 8-6
USE OF MATERIALS

	MATERIALS USED	TIME INTERVAL, MINUTES					
		5	10	15	20	25	30
USED BY THE TEACHER	Textbook						
	Chalkboard						
	Audiovisual aids						
	Plants for demonstration						
	Animals for demonstration						
	Inanimate objects for demonstration						
	Models for demonstration						
	Data-recording sheets						
	Wall charts						
	Trade book(s)						
	Other						
USED BY THE STUDENTS	Textbook						
	Chalkboard						
	Audiovisual aids						
	Plants to manipulate						
	Animals to manipulate						
	Inanimate objects to manipulate						
	Models to manipulate						
	Data-recording sheets						
	Wall charts						
	Trade book(s)						
	Other						

suggestions. You could develop checklists for other topics—such as evidence of creativity, management techniques, signs of disequilibrium versus equilibrium, supportive or nonsupportive behaviors, or use of the thinking operations.

The checklist technique is limited by the number of specifics included. If you wish to develop an observing instrument which allows for greater flexibility, the next technique—coding—will be helpful.

Coding Technique

The observation form presented in Table 3-3 (page 60) for recording the results of Piagetian-type interviews is a combination of a descriptive record and a coding technique. Coding systems were used to categorize (1) children's responses to the conservation tasks (more, less, same); (2) the types of reasoning they demonstrated in responding to the tasks (logical necessity, compensation, reversibility, centering on the object, centering on the action, or no basis); and (3) their general reactions to the interview questions (random, romancing, suggested conviction, liberated conviction, or spontaneous).

Table 8-7 is an example of another coding system for observing types of reactions to questions in general. Underline the T if the question originated with the teacher; underline the S if it originated with the student. If possible, record the actual question asked or at least some sort of descriptive summary that can be referred to later. The following key explains the codes for reactions

Ra: random, noncommittal
Ro: romancing or fantasizing
Su: suggested conviction (answer made to please the questioner)
L: liberated conviction, the sharing of one's best thoughts
Sp: spontaneous conviction or the recall of previous ideas
NR: no response after several seconds
QR: question needed to be rephrased
QQ: question elicited another question

(Other categories could be added or substituted according to your special interests. This is true for all the examples of observing instruments presented.)

When interpreting the data on reactions to questions, any or all of the following questions could be considered:

What is the rank order of the frequency of the types of reactions?

Which questions elicited the most reactions?

Table 8-7
OBSERVATIONS OF REACTIONS TO QUESTIONS

QUESTIONS	Ra	Ro	Su	L	Sp	NR	QR	QQ
(T) (S) 1.								
(T) (S) 2.								

Which questions most often resulted in no response?

If students directed questions to the teacher, how did the teacher react?

To what extent were questions rephrased when an initial reaction was a suggested conviction or no response?

Observing and Classifying Classroom Questions

Regardless of whether you decide to use a descriptive record, a checklist, or a coding technique to observe classroom interactions, most likely you would eventually observe that over half of the instructional time available (even as much as 70 to 80 percent) is spent on asking and answering questions. Some studies indicate that the average number of questions asked per minute is 3 to 6—or between 350 to 450 questions per day. (Read Roger Cunningham's chapter, "Developing Question-Asking Skills," in *Developing Teacher Competencies*, 1971, pp. 81–130.) Because questioning is so important compared with the general considerations presented so far, it merits special attention as a specific example of observing and interpreting a type of classroom interaction.

Questions are very much like keys. If they are applied carefully to the correct lock, they can free all kinds of facts, feelings, reactions, ideas, opinions, and even additional questions. When they are applied carelessly, they might have little or no effect, or even a negative effect—a "jammed lock." Recall your conversations with children when you were using the Piagetian-type interviews or tasks. Some questions might have elicited only a blank stare or a mumbled "I don't know." Others might have caused the child to become visibly uncomfortable, nervous, or embarrassed. Still others might have stimulated an avalanche of words or an eruption of original ideas.

Ever since the origin of formal education, questions have probably been asked for the same general purposes: to find out what students know or don't know, and to review and summarize previous learning. But there are some other purposes which are less obvious and, in many ways, more important. Questions can be used to motivate, to arouse interest, and even to stimulate students to make their own discoveries. ("Why does this bulb shine more brightly than the other? How can we find out?" Or, "What would you most like to learn about these rock samples?") Some questions can also help the learners to consider new relationships. ("In what ways is a green plant like a factory?") If the teacher is a real inquirer, he or she can serve as a model for developing a questioning attitude in the students. ("When I was walking this weekend, I found this strange object which has really puzzled me. What do you think it is? How do you think it works? Where do you think it came from originally? What's the strangest object you've ever wondered about?")

Unfortunately, questions can also be used in negative ways. When students are confronted with a question like "What was *the* most important thing we learned about flowers yesterday?" they are forced into a game of mind reading. This kind of question deals more with what the teacher wants to hear than what the students want or need to say. The teacher who asks an inattentive child, "What did I just say about the life cycle of a moth?" is using the question as a disciplinary technique. When questions are used in these ways, they are misuses of communication. As such, they can damage rather than build up the self-concept. They do not promote real interpersonal sharing, the essence of communicating.

The most important purpose of classroom questioning is probably twofold: to promote thinking and to *request* (not demand) the sharing of one's thoughts. Interrogatives and imperatives differ in more ways than the final punctuation mark. A question is an *invitation* to share, with the notion of compliance or noncompliance implied. An imperative is a command; the expectation for a specific performance is explicit.

What is so extremely powerful and yet also so precarious about questioning is that one can hope for just about any sort of shared response, but there is no guarantee that the desired answer will be given or recieved. If you ask for trivia, you may or may not receive trivial answers. The same is true of questions dealing with the higher levels of thinking. But why, then, should we consider questioning an important aspect of classroom interactions?

The answer seems to depend on a belief in teaching and learning as an interpersonal relationship based on trust, respect, and caring. If a teacher uses questions deliberately and sincerely to guide children's thinking (rather than as a personal ego trip or as a mechanical method to avoid silence or unwanted talk)—and if the children sense, respect, and accept these invitations—there are no limits to the potential for meaningful sharing. The secret of success is for the questioner (who can be either the teacher or the students) *to become more fully conscious of the reasons for asking the questions.* With this consciousness you will be in a much better position to ask the kinds of questions most likely to elicit what you hope to discover. If questions are indeed keys to unlocking others' thoughts, feelings, or actions, consciousness of the questioner's intentions gives substance and shape to these keys.

Types of Questions

It's fairly easy to accept the notion that understanding exists in varying degrees. For example, the kind of "understanding" which is demonstrated simply to pass a test and then quickly forgotten is entirely different from understanding which can be applied to solve a new problem, modified to make a personal statement, or judged against related materials. (If more examples would help, examine *Taxonomy of Educational Objectives, Handbook I: Cognitive Domain*, Bloom, 1967.) The notion that questions can be phrased deliberately to elicit various levels of understanding is also somewhat easy to accept but more difficult to apply. Some examples of categories of questions are needed.

Sanders's System Norris Sanders based his questioning categories on a slight modification of Bloom's levels of understanding. In his book *Classroom Questions: What Kinds?* (1966), he demonstrated how questions can be deliberately phrased to elicit a particular level of thinking, especially in the area of social studies. The following examples show how his categories can be geared to science topics:

Memory: How many body segments does a mealworm have?

Translation: How could the leg movements of a mealworm be mimicked?

Interpretation: In what ways are a mealworm and a caterpillar similar?

Application: What will cause a snail to move faster?

Analysis: Which stimuli cause a mealworm to back up?

Synthesis: How might you design a "Mealworm Marathon"?

Evaluation: What are the best conditions for studying the behavior of mealworms?

If you're interested in a more thorough explanation of Bloom's taxonomy applied to questioning, read Francis Hunkins's book *Questioning Strategies and Techniques* (1972). Hunkins also stresses the inherent value of the cognitive taxonomy as a questioning strategy. If you wish to help children develop higher levels of thinking at the analysis level, for example, questions should first focus on the knowledge level, then on the translation level, then on the interpertation level. Since the higher levels of understanding subsume the lower levels, the taxonomy can serve as a guide for reaching the intended objective. But as is often true of traveling in general, there is usually more than one route available.

Aschner-Gallagher System   The Aschner-Gallagher system is one of the most widely used systems for categorizing questions. It classifies questions first as "narrow" ("closed") or "broad" ("open") and then classifies "narrow" questions as "cognitive-memory" or "convergent" and "broad" questions as "divergent" or "evaluative." Roger Cunningham (1971), Arthur Carin and Robert Sund (1971), and Patricia Blosser (1973) have utilized it extensively in their writings and research. A summary of the main categories of the system, along with a few examples of questions, is given in Table 8-8.

In her monograph *How to Ask the Right Questions*, Patricia Blosser adds two categories to the four shown in Table 8-8 (cognitive-memory, convergent, divergent, and evaluative); these are the *managerial* and *rhetorical* (1975, pp. 2–3).* Managerial and rhetorical questions deal less with levels of thinking and more with simply operating a classroom. The purpose of managerial questions is to keep the classroom action moving toward the learning goals. Some examples are: "Who will be in charge of gathering the seeds?" "Which team wants to report its observations first?" "Can you change the hamster's bedding before recess?" Rhetorical questions are used mainly for the sake of emphasis or reinforcement. Many times no answer is expected, or the answer is implied by the question itself. For example, "That movie showed the sun as the source of all energy, right?" "Wasn't that an interesting result?" "What about life on other planets?"

The Aschner-Gallagher system can easily be arranged into a checklist and coding system for observing and classifying questions, as shown in Table 8-9.

*Cited by permission of Patricia Blosser.

Table 8–8

OVERVIEW OF ASCHNER-GALLAGHER CATEGORIES OF QUESTIONS

NARROW OR CLOSED QUESTIONS Require recall ability rather than reasoning. Expected answers tend to be shorter, predictable, of the "one right answer" variety.	**COGNITIVE-MEMORY**	*Possible cognitive behaviors to be elicited:* Name, identify, designate, describe, define by rote, choose affirmatively or negatively (answer "yes" or "no"), and, in general, give any evidence of recall or the ability to memorize. *Examples:* What is the name of this plant? How many body parts does an insect have? Can aluminum foil conduct electricity?
	CONVERGENT	*Possible cognitive behaviors to be elicited:* Identify relationships, explain, compare, contrast, classify. *Examples:* How do fish survive during freezing weather? Why do roots usually grow downward? What's the difference between an observation and an inference?
BROAD OR OPEN QUESTIONS Require higher levels of thinking. Expected answers tend to be longer, unpredictable, varied and open to debate.	**DIVERGENT**	*Possible cognitive behaviors to be elicited:* Create, synthesize, infer, predict, hypothesize, reconstruct, imply, identify new patterns or forms. *Examples:* In what ways can the problem of pollution be lessened? If you could genetically engineer the fastest animal, how might it look? How would you prepare for living on the moon?
	EVALUATIVE	*Possible cognitive behaviors to be elicited:* Develop criteria, judge, defend a position, justify a choice, evaluate. *Examples:* Which materials make the best magnets? Which interpretation is the most accurate? How can we make better use of our natural resources?

The recording of observations can be indicated by tallies. Analysis of the data could include the percentages of the types of questions asked and the distribution of the questions over time. You might look for any evidence of a sequence of questions—such as more cognitive-memory questions at the beginning of a lesson, followed by more convergent questions, with divergent and evaluative questions occurring more frequently toward the end of the lesson. If such a sequence was observed, would you infer that a deliberate questioning strategy was used or that the pattern was accidental or unplanned? On what would you base your judgment?

The data from this sort of classroom observing could be very helpful to a teacher who values classroom interaction or dialogue rather than monologue. Suppose that the only questions from students observed during a 30-minute science lesson were of the managerial type—"When is our observation report

Table 8-9
OBSERVATIONS OF CLASSROOM QUESTIONS

TYPE OF QUESTIONS		INTERVAL, MINUTES							
		5	10	15	20	25	30	35	40
Cognitive-Memory	T								
	S								
Convergent	T								
	S								
Divergent	T								
	S								
Evaluative	T								
	S								
Managerial	T								
	S								
Rhetorical	T								
	S								

due?'' ''Where should I put the rock samples?'' Obviously the existing strategy (or lack of strategy) is ineffective, at least as regards developing a questioning attitude among students. (Francis Hunkins's book *Involving Students in Questioning*, 1976, might be quite helpful if this is the case. He presents numerous techniques for increasing students' involvement in learning and social participation, which can help both the teacher and the students to become more active questioners.) Or perhaps the teacher needs to become more aware of the types of reactions resulting from the classroom questions.

The more you observe classroom questions and questioning strategies, the more you will discover the difficulty of categorizing questions—especially if the context in which a question is asked is unknown. Perhaps the most important benefit that can be derived from using any system of categorizing questions is an increase in awareness. *The more conscious you are of your questioning, the greater the chances of phrasing questions so that they will elicit what you're after.* By analyzing classroom questions (or even those printed in science textbooks), you can also acquire some clues on effective and ineffective questioning techniques. Compare your discoveries with the discussion in the next two sections.

Ineffective Questions and Questioning Techniques

As was just suggested, if you've already attempted to observe and analyze classroom questions, it would not be surprising if you found them difficult to categorize. You might have also discovered that it is difficult to judge the effectiveness or ineffectiveness of questions on their face value alone. Some teachers can ask seemingly poor questions yet, because the students are highly interested or cooperative, receive excellent responses. The opposite can also occur. The following types of questions, however, are generally ineffective and therefore should be avoided.

Questions which call for only "yes" or "no" answers. Seldom is a teacher interested in simply a positive or a negative response. Most of the time this sort of question is followed up with a "Why?" or "What's the reason for your choice?" or "How come?" If it is an explanation which is being sought, the "yes-or-no" question is a waste of time. (It is also an invitation to chance guessing.) The best way to avoid futility is to identify the "yes-no" question before you say it. The identification is simple. Any question which begins with an auxiliary verb (a "helping verb"—*is, are, were, does, do, could, should, would, might, shall, will, have, has, had . . .*) automatically asks for a "yes-no" response. But identifying this type of question beforehand is not all that simple—considerable effort is needed to think before questioning.

Questions which contain difficult or unfamiliar words. The purpose of any form of communication is sharing. Words, and their organization into statements or questions, have only as much meaning as the participants in the communication process bring to them. Therefore, it is not only ineffective but also unfair to formuiate questions containing unfamiliar and unduly difficult words. Suppose a group of first-graders has been placing objects on a balance. The question, "How does the position of the balance beam indicate the equality of gravitational force between these two specific material objects?" most likely would be meaningless and ineffective. The question, "How do you know these two objects weigh the same?" would make much more sense.

Vague questions. When questions such as "What can you tell me about animals?" or "How do you feel about rocks?" or "What about electricity?" are asked, children tend to respond with blank expressions and silence. Most students, even the very young ones, will realize that the question is impossible to answer. They'll simply wait for the questioner to make the same discovery and hope that a more specific question is eventually offered.

Statements which begin as declaratives and turn into interrogatives. For example, "Yesterday we learned that there are three basic types of rocks; these were . . . what?" or "That part of a leaf which is most like the lungs in our body is . . . where?" Statements converted into questions are unfair because they lull the listener into passivity and then suddenly call for action. Usually both the students and the teacher lose. The students are caught off guard, and the teacher must retreat and rephrase a more effective question.

Reliance on any of these ineffective questions is, of course, a *poor questioning technique.* But there are additional ways that questioning techniques can be ineffective. Perhaps you have experienced some of the following poor questioning techniques during your own educational career.

"*One right answer.*" For the bright student, at least initially, this technique may be an enjoyable way to show off. But it soon becomes tiresome because of the limited thinking required. For the not-so-bright learner or for one whose memory is not always in high gear, this technique can prove to be a dead end. Imagine the mounting frustration of the child who never (or, at best, seldom) knows the right answer, and the resultant steady decline of self-confidence and perhaps even of self-concept.

Machine-gun approach. This technique consists of a rapid-fire, nonstop series of questions. For example, after the pendulum activity ("Programing Pendulums," page 169), a teacher might shoot off a round of questions: "What caused the pendulum to swing faster? Did you consider all the important variables? Did you construct a graph? What data did you include in your graph? What effect did changing the mass of the bob have on the frequency?" Most students would probably sit silent, waiting for a single on-target question or one within their grasp.

Speeding. This technique is similar to the machine-gun approach but does not necessarily involve a volley of questions. It implies moving too quickly either from one respondent to the next or from one question to an entirely new one. Unfortunately, questions in the classroom are sometimes handled in the same way as the superficial, fleeting "Hello, how are you?" Perhaps *no* question would be better than one which suggests indifference or impatience.

"Everything is beautiful." Lack of interest and lack of discernment are evident when the questioner accepts any and every response without comment or without any request for clarification. Sometimes a teacher uses this technique because he or she is thinking of the next question—anticipating instead of concentrating on the present.

Senseless sequencing. This implies asking questions haphazardly rather than deliberately attempting to guide thinking from lower to higher levels of involvement. It also implies failing to relate the next question to previous responses.

These are only a few of numerous possible ineffective questioning techniques. What other examples have you experienced? As you ponder this question, perhaps you will also become more mindful of some of the positive questioning practices.

Improving the Skill of Questioning

The time spent in asking and answering a single question may sometimes amount to only a few seconds. But the time and effort needed to use questions effectively is considerable. In order to give you a better picture of the process of questioning, I will present it as a sequence in chronological order.

Before the Questioning Try to plan your key questions in advance, either mentally or—better yet—by jotting them down so that you can actually see them. When you examine your planned questions, check for variety and sequences. Try to include questions which deal not only with cognition but also with the process skills and with attitudes. Also, think about possible strategies for sequencing your questions from low levels to higher levels. Usually a great many questions deal with *who, what, when,* or *where.* Plan to follow up these sorts of narrow, recall questions with slightly broader questions pertaining to *why* or *how.* Then try to determine the real strength of the respondent's convictions with questions like "What's the basis for your answer?" or "What are you really saying?" Or use the very natural sequence of following a divergent question with a related evaluative question.

It will also be helpful to realize ahead of time that not only how you phrase your questions but also how you *emphasize* certain words makes a big difference. Consider the following forms of the same question and determine (or at least imagine) the different meanings implied by the various emphases:

How might you test that hypotheses?
How *might* you test that hypothesis?
How might *you* test that hypothesis?
How might you *test* that hypothesis?
How might you test *that* hypothesis?
How might you test that *hypothesis?*

Also, give some advance thought to ways of preparing your students for some of your higher-level questions. For example, "My next question might be a challenge, so think about it before you try to answer," or "You didn't read about this question, but perhaps you've experienced something related to it elsewhere." During the planning phase for classroom questioning, try to picture mentally what you hope to accomplish. How well do your key questions relate to your intentions or expectations? Keep in mind the primary purpose of your questions—that is, an invitation for sharing, not to an inquisition.

During the Questioning   Always ask your question first, before calling upon an individual to respond. The opposite approach captures the attention of only one person while the rest of the class can remain passive or free to wander mentally. Also, try to avoid any questions which could invite mass shouting. ("How many of you can describe these objects for me?" "Which group wants to be in charge of sampling the pond water today?")

Try to direct your questions to a variety of students. Too often teachers tend to direct their questions to the most vociferous children or the most visible hand wavers, or to those situated in the more central section of the room. On the basis of an analysis of classroom interactions by Raymond Adams and Bruce Biddle (1970, p. 49), it can be estimated that 63 percent of students' responses come from those seated in the "teacher's inverted T" (diagramed in Figure 8-1). One way to increase students' participation is simply to move about in the room from time to time. Another way to involve more students in responding to questions (since only one should speak at a time) is to call on additional students to either agree, disagree, or modify a given response. Try to avoid the easy habit of

Figure 8-1 Of this class, 63 percent is participating.

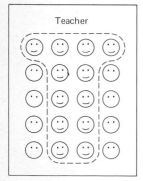

directing the more difficult questions to the bright students, simpler questions to the less capable, and disciplinary questions to the troublemakers.

One of the most important ways to improve your questioning skill in the classroom is simply to *slow down.* Perhaps you recall that this same suggestion was given when the implications of Piaget's ideas were discussed, but it's certainly important enough to be repeated. In general, the more difficult the question, the greater the need for "think time" for both the questioner and the answerer. Why is this so? What generally goes on in one's head when one is attempting to answer a higher-level question? (This is not intended as a rhetorical question. Think about your own thought processes in attempting to answer it. Then compare your results with the next comments.) First, you must analyze the information being requested and make a judgment. That is, how does what you know mesh with what is being asked? This information must then be mentally synthesized and quickly prepared for verbalization—that is, *if* another question can be answered affirmatively: Are you willing to take the risk? For verbalization entails success or failure, acceptance of the answer or possible rejection. Maybe you've never analyzed these steps in detail, but they do tend to indicate fairly accurately what is required to answer a higher-level question. Yet teachers generally allow a student only 1 or 2 seconds to answer any question before calling on another student or formulating another question or comment. The human brain has often been compared to a fantastic computer. But few adults—and far fewer elementary school children—can race through the steps involved in answering in a few seconds. Therefore: *slow down*—especially when you ask a higher-level question. You might be pleasantly surprised by the results for both you and your students. (If you're interested in learning more about the effects of slowing down the questioning pace and increasing "wait time," read pp. 273–295 in Mary Budd Rowe's *Teaching Science as Continuous Inquiry,* 1978.)

During the Response   At this point your attention should be directed entirely toward respect for the child's response. It's all too easy to ask questions which lead to comments we want to hear rather than to what the learners need or want to say. If the students view the teacher as a sovereign or sage, variations of suggested conviction will dominate the scene rather than liberated, independent, honest thinking. How can you best convince your students that you are most interested in *their* thoughts and not in the answers that they think might please you? Perhaps the basis for the answer to this question is to be found in the next specific example of an important classroom interaction—listening.

Analyzing the Skill of Listening

This is one of the most neglected aspects of the skill of observing and, of course, communicating. Most research on listening has been done only since the 1950s. Yet we spend much time listening—and the amount probably increases as you go up the educational ladder. Elementary school children spend a great deal of their school time (probably over 50 percent) simply listening; high school and college students spend perhaps as much as 80 to 90 percent. Nichols (1960) and other researchers have shown that untrained listeners retain only about half of what they hear—and half of this is lost after several months. This means that we normally operate at about 25 to 30 percent of listening efficiency. It's also fairly predictable that the percentage decreases even more as the length of time required to listen increases, and also as we grow older. Why are we such inefficient listeners? All but a few of us (perhaps 3 to 5 percent) are born with the ability to respond to the sensation of sound—that is, with hearing ability. But *listening* (unlike *hearing*) is not an inherent capacity. Listening involves being able to perceive and interpret differences among sounds. Why have we not developed this ability more keenly?

Some Problems Related to Listening

If a tape recording of your past few days were to be replayed, how many surprises do you think you might discover? (Did I say that? Is that what she really said? I had no idea he was talking about that.) What did you miss the first time? What might have been the reasons there for the misses? Most likely if twenty-five people heard such a recording, there would be twenty-five different problems related to listening. There's no way of accurately knowing all the causes of ineffective or inefficient listening; but some similarities and generalities can be gleaned from varied experiences. These can be classified in three major categories: physiological and mechanical problems, unconscious habits, and barriers created by personal bias.

Physiological and Mechanical Problems One of the basic problems of listening is coping with *speed differentials.* We normally speak at a rate of about 125 to 150 words per minute. Someone delivering a speech, lecture, sermon, or the like usually slows down this rate to approximately 100 words per minute. But the human brain is capable of thinking or reading many more thoughts or words per minute—perhaps anywhere from 400 to 500 or even more. We must operate with two very different rates—the *actual* rate being spoken and our *capacity.* The differential between the speaker's output speed and the listener's input capacity creates a difficulty which can be overcome only by deliberate efforts on the part of the listener to arrest or control the mental processing capacity and attend to the sound stimuli from the speaker.

A somewhat similar problem is *stimulus bombardment;* this is due to a variety of surrounding distractions. Have you ever tried listening to a friend on the telephone with the radio on, the dog barking, the oven timer ringing, and an airplane passing overhead? Try listening to a child's response to a question when others are talking, the pencil sharpener is being used, furniture is being moved, the second-graders are having a relay race near your windows, and the instrumental music class next door is warming up. Then you'll realize the difficulty caused by stimulus bombardment. Except for some rare moments in the stillness of the night, there are very few times when many sounds are not present. The greater the number and volume of distractions, the more difficult it is to listen, and the problem is compounded in a room where no provision has been made for sounds to be absorbed.

Another difficulty in the general category of physiological and mechanical problems is simply *fatigue*. The comment "I'm too tired to listen" can be a very honest and accurate one. When you are attentively listening, there is a slight rise in your body temperature as well as an increase in your pulse rate. In other words, listening requires an expenditure of physical energy as well as mental involvement. If you want to listen attentively to students' responses to your questions, especially to the higher-level ones, don't make a habit of saving such questions until the end of the school day.

Unconscious Habits   It's easy to develop poor listening habits without realizing that they've become part of your behavior patterns. Listening requires full attention, concentration, and a good deal of energy and effort. Because of training, the example of others, the ingrained rule "Pay attention," and the desire to please by at least appearing to be interested, it is common to *feign attentiveness*. A chin-in-hand, forward-leaning posture and piercing eyes do not always indicate effective listening. Even with this disguise, one can be day-dreaming or fighting off sleep during a boring presentation. Most of us have probably been on both the giving and the receiving end of faked attentiveness at some time in our lives. It can become a serious problem when we are unaware of it. You might try thinking about the relationship between this phenomenon and false accommodation.

Another unconscious habit which hinders good listening stems from *impatience*. Sometimes we "jump the gun," that is, we anticipate what we expect to hear and fail to listen to what is actually being said. Sometimes we concentrate on just one idea and tune out the rest of what is being said. When impatience becomes uncontrollable, interruptions occur. If you've ever tried to talk seriously with an interrupter, you know how your own patience can be tested. Interrupters who are oblivious of the habit can make real communication virtually impossible.

The use of *instant replays* on television has also become a hindrance to observing, communicating, and, of course, listening skills. More and more we've become conditioned to attend rather haphazardly at first and wait for the reply. Unfortunately, teachers frequently promote this expectation by consistently repeating each child's comments. This can become an unconscious habit which is very difficult to change.

Because most of us are far more accustomed to the presence of sounds than to their absence, we can develop a habit of *fearing silence* or at least being relatively uncomfortable with it. This can cause us to want to quickly fill up the void with a new array of sounds or words. This drastically reduces listening and absorption time. Even though sounds can be received by the brain extremely rapidly, deciphering and comprehending a message requires more time than is usually given. It's something of a paradox that teachers often complain about noisy classrooms or overly talkative children and yet avoid silence themselves when questioning and listening are involved. Waiting and being still are part of listening and comprehending.

Barriers Created by Personal Bias   Just about any kind of prejudice can make effective listening more difficult. Some listeners overreact to certain words or labels—like *abortion*, *taxes*, *strikes*, *freedom*, *leftist*, or *freaks*. Some words can stimulate such violent vibrations that few of the speaker's subsequent words can be heard until peace and calm are restored within the listener. Sometimes, too, listeners fail to listen because they dislike the speaker's accent, intonation, style or pace of delivery, mannerisms, style of clothes, or any number of possible idiocyncrasies. These are irrational reactions.

The know-it-all type of person is also generally a poor listener, though perhaps not quite so irrational. For this person pride, rather than emotion, is the major obstruction. Comments like "I've heard all this before" or "Here we go again; I bet this will be a real bore" or "There's nothing new to be said about . . ." are indicators of the know-it-all's impenetrable bias.

A weaker problem related to listening deals with one's own preferences (rather than prejudices). Some people, especially those who enjoy the printed word, simply prefer reading to listening. When this sort of person becomes a captive audience, various forms of antagonism or irritation might exist. It's possible that a preference for reading could develop into a prejudice against listening.

At this point you may be feeling that there are numerous problems related to listening—even more than have been described in this brief section. Are we destined to fight a losing battle? Can any of these problems be overcome or lessened?

Improving the Skill of Listening

In their programed text *Study Skills: A Student's Guide for Survival*, Carman and Adams (1972, pp. 7–17) suggest the acronym LISAN as a device to improve the skill of listening. Although their plan is directed more toward improving lecture-type listening, it can readily be adapted to observing and communicating in the classroom. The following is a summary of their suggestions:

L = lead. Don't simply try to follow what is being said. Try to anticipate what you think is going to be said. In other words, you'll hear more and learn more by being active rather than passive.

I = ideas. Look for the main ones. Try to analyze what is being said. Determine what is important and what is superficial; what is fact and what is fiction; what is subjective and what is objective.

S = signals. Recognize them when they occur. Often a speaker will highlight certain points. (For example, "These are the *four types* of erosion we'll be examining. . . .") Signals can help you distinguish the main ideas from the superficial ones.

A = active (rather than passive) *involvement.* Let the speaker know through your *eye* contact that you are attending to the message. When it is appropriate (and honest) to do so, ask questions or seek additional explanations.

N = notes. Take notes physically if you can, or at least mentally. Then try to organize or classify them, interpret them, and summarize them.

Nichols (1960) makes similar suggestions, but he includes some reinforcing ideas. Consistently attempt to identify the speaker's supporting material or references. This will greatly increase your active involvement in the listening process. Identifying *why* something is said may often be more important than what is actually said. At every convenient opportunity—for example, a slight pause—mentally summarize the speaker's main point or points. Use the phenomenon of speed differential to your advantage. While the speaker's words are being said at their laboriously slow rate, use your swift thought-processing ability to recapitulate the message given so far. When training children to read orally, we encourage them to look ahead at the coming words. When training yourself to improve your listening ability, try to be tuned in to past, present, and future words.

Another strategy that can be employed is like a self-imposed dare. Challenge yourself to determine how much of what a speaker says can be of use to you—if not immediately, perhaps in the future. Or take on a different sort of challenge. Generally, we tend to listen for that which proves us right. Try listening for what proves you wrong. Or you might reexamine each of the problems related to listening which were previously described and develop a counterattack for each. Most important, examine your own *attitude* toward listening, toward observing, and toward communicating in general. Your attitude probably has a greater influence on your behavior than any other factor.

Establishing Some Bases for Decisions

Observing the classroom interactions which occur during science encounters—especially if it is done systematically on the basis of your personal choices of *what*, *why*, and *how*—offers you limitless learning opportunities. There are an infinite number of factors that could be identified, observed, and recorded by means of your own observation instruments, or instruments developed by others, or a combination of both. The benefits that can be derived from your classroom observing will greatly depend on your intentions and your degree of interest. The ultimate responsibility for any of your learning encounters resides with you. This is particularly true of the information and suggestions presented in this chapter. As was explained before, it takes much experience before professionals can fully apply their skills—hours, days, months, and even years of observing, classifying, measuring, analyzing, interpreting others' behavior and their own responses. If the information and suggestions in this chapter are to be of use, you must *try* them. (The word *practical* is derived from the Greek *prassein*, "to do.")

Implementing an Observing Instrument

Spend some time now to formulate your own ideas for classroom observing of science with children. What do you wish to observe? What will be most helpful? Why do you wish to observe this? Why is this worth observing? How will you observe what you've selected? Which observing technique—descriptive record, frequency record, or coding—will facilitate the most practical and meaningful record of observations?

Develop your plan and implement it. What are your results?

Try to keep both your observing instrument and your implementation strategy simple and manageable. You will find it less cumbersome and perhaps more meaningful to begin by observing just one particular strand of behavior rather than attempting to unravel the entire complex web of classroom action. Develop or select clear, objective, nonevaluative descriptors for the categories to be observed. If possible, observe a variety of settings. Classroom observing, like the observing of material objects or natural events, requires repeated practice before any substantial degree of proficiency can be achieved.

Analyzing the Results

It's impossible for me to predict accurately what will result from your systematically observing science with children in the classroom. But the following possibilities should begin to make more sense, and the questions related to them should become more answerable, in proportion to the quantity and quality of your experiences with classroom observing.

Analysis of the results of your classroom observing can provide meaningful bases for improving your future instructional decisions. More specifically, analysis of classroom interactions can increase your self-awareness, particularly in terms of self-confrontation and self-identity. These two aspects of self-awareness are easy to avoid but most worth considering.

Self-Confrontation How long has it been since you were an elementary student, on the other side of the desk? How does your frame of reference as an elementary student compare with what you have recently observed in elementary classrooms?

In what ways have your perceptions of the teaching and learning processes been reinforced, confused, or modified as a result of your classroom observing? Considering the science with children that you've observed, would you say that science is being presented as a noun, as a verb, or as both?

How do you want to function as a teacher? How do you not want to function? Do you want to be a trainer or a facilitator? Do you want to develop children's ability to memorize or their ability to reason? Do you want to show and tell children all there is to know or allow them to invent and discover for themselves? What's the basis for these choices?

What kind of classroom atmosphere do you want to maintain or at least attempt to promote? Warmness or coldness? Trust or distrust? Suggesting or demanding? Accepting or taking? "Please touch" or "hands off"? Community building or guerrilla warfare?

As you observed, to what extent were you aware of individual differences among children? In what ways did the children you observed demonstrate different approaches to solving problems, different levels of understanding, different degrees of self-motivation, different abilities to think and behave independently? How did individuals respond to influences of the peer group as opposed to the teacher's expectations?

When did disequilibrium occur? How was it handled by the teacher and the children? What evidence did you observe of meaingful accommodation and false accommodation? What happened to the children on the dead-end route? What might have affected their choice? How frequent were suggested convictions as compared with liberated reactions? What changes might you have introduced to try to guide the children toward more meaningful interactions?

To what extent were you aware of the use of time and materials by the teacher and the students? How much time did the teacher spend in showing and directing as opposed to leading and suggesting? In correcting negative behavior as opposed to encouraging positive behavior? How much time did the students spend listening to the teacher or listening to each other? In talking (with or without permission) to the teacher or to each other? How much time was spent on lesson-related actions and how much on distractions? What types of materials were available, and how were they used or abused?

These are just a sample of questions for self-confrontation. Which have become more answerable as you analyze your observations of children interacting with science in classrooms? Chapters 9 and 10 will give you more opportunities to search for answers. So might the next facet to be considered—self-identity.

Self-Identity   One of the least thought-about yet most significant aspects of observing in a classroom is the phenomenon of "playing teacher." What is the basis for the behavior of children "playing teacher"? For those players who are too young to have attended school, chances are that impressions have been picked up from older siblings, playmates, or parents, who in turn have taken their impressions from experiences with teachers. Watching such role playing can be both funny and serious. I once observed a second-grader "playing war" with some of his friends. He assumed the role of a tank driver in the heat of a battle and was very convincing. After the group grew tired of playing the war game, they decided to play school. The former tank driver became the teacher; and there were few differences between his actions as a tank driver and as a teacher. (Previously he had been a very docile victim in an outer-space game, so his behavior could not be categorized totally as aggressive.) Later, I asked him to describe his teacher. The teacher's last name wasn't Tank, but that would have been fitting if the boy's description was accurate. I could describe instances of transformations some student teachers have undergone after a relatively short period of internship. Some veteran teachers also exhibit unaccountable personality differences between the teacher's lounge and their classrooms.

The comment is often made in educational circles, "We teach the way we were taught." If such a generalization is true, the results may vary from magnificent to inconsequential to disastrous. Regardless of your past educational experiences, satisfying or dissatisfying; regardless of the number of times you have heard or will hear, "Forget all that educational theory. In the real classroom, we do it this way"—*you* must decide for yourself the kind of educator you want to be. Common sense as well as research indicate that effective teachers generally have an accurate, positive self-concept; are flexible; can communicate clearly; and are perceived as fully human. That is, they can laugh, empathize, become justly irritated, and relate warmly and naturally to students and to others. (See Don Hamachek's article, "Characteristics of Good Teachers and Implications for Teacher Education," 1969, pp. 341–345.) How well does (or will) this description apply to you? Classroom observing can provide you with some real bases for making your decision. On the basis of the science with children that you have observed, how would you expect the students to "play teacher"? If they had *you* presenting science to them, what would their imitations be like?

Additional Considerations on Observing Classroom Interactions

This chapter can never be considered finished, just as the need for the skill of observing in general can never be removed. It is difficult to predict how many daily instructional decisions will depend on your classroom observing skill. Will awareness stem from your attitude, or will your attitude enhance your awareness? Perhaps attitude and awareness are too intertwined to be distinguishable. The vital point is that you increasingly care about both when you provide, observe, and evaluate classroom interactions.

Encounters

1. Summarize your experiences of observing in a classroom. Describe what you wanted to observe, the reasons for your choice or choices, the means you used to record and categorize your observations, and your results. Explain how your observing experiences were, or were not, beneficial.
2. If *you* had been teaching the class you observed, what would you have done similarly or differently? What are the bases for these decisions?
3. One of the major results that can be derived from observing classroom interactions is an increase in awareness. In one section of this chapter, numerous questions were presented (pages 238 to 239). Which of these questions do you consider most important in terms of bases for decisions about self-improvement?
4. Select any activity you completed in Part Two and design a series of questions related to it. Construct at least three questions of each type

discussed in this chapter: cognitive-memory, convergent, divergent, and evaluative.

5. On the basis of what you've experienced, what do you consider to be the most needed improvements for classroom questioning? What are the bases for your decisions?

6. Compare the types of questions presented in two or three different elementary science textbooks. (If possible, select similar topics from textbooks for the same grade level.) What are the percentages of cognitive-memory, convergent, divergent, and evaluative questions? What do you consider to be the effective and ineffective uses of the questions you examined?

7. Why might there be a high correlation between false accommodation and feigned attentiveness? What other problems hinder effective listening?

8. As a result of your interacting with the ideas in this chapter, in what ways do you want to change or retain some basic observational skills needed to do science with children?

9. Your choice.

References

Adams, Raymond S., and Bruce J. Biddle: *Realities of Teaching*, Holt, Rinehart, and Winston, New York, 1970.

Bloom, Benjamin S. (ed.): *Taxonomy of Educational Objectives, Handbook I: Cognitive Domain*, McKay, New York, 1967.

Blosser, Patricia E.: *Handbook of Effective Questioning Techniques*, Education Associates, Worthington, Ohio, 1973.

————: *How to Ask the Right Questions*, National Science Teachers Association, Washington, D.C., 1975.

Carin, Arthur A., and Robert B. Sund: *Developing Questioning Techniques: A Self-Concept Approach*, Merrill, Columbus, Ohio, 1971.

Carman, Robert A., and W. Royce Adams, Jr.: *Study Skills: A Student's Guide for Survival*, Wiley, New York, 1972.

Cunningham, Roger T.: "Developing Question-Asking Skills," in James Weigand (ed.), *Developing Teacher Competencies*, Prentice-Hall, Englewood Cliffs, N.J., 1971.

Hamachek, Don: "Characteristics of Good Teachers and Implications for Teacher Education," *Phi Delta Kappan*, February 1969.

Hunkins, Francis P.: *Questioning Strategies and Techniques*, Allyn and Bacon, Boston, Mass., 1972.

————: *Involving Students in Questioning*, Allyn and Bacon, Boston, Mass., 1976.

Nichols, Ralph G.: "What Can Be Done about Listening," *The Supervisor's Notebook*, vol. 22, no. 1, Spring 1960.

Rowe, Mary Budd: *Teaching Science as Continuous Inquiry*, 2d ed., McGraw-Hill Book Company, New York, 1978.

Sanders, Norris M.: *Classroom Questions: What Kinds?* Harper and Row, New York, 1966.
Simon, Anita, and E. Gil Boyer (eds.): *Mirrors for Behavior III: An Anthology of Observation Instruments*, Research for Better Schools, Philadelphia, 1974.

Suggested Readings

Cartwright, Carol A., and G. Phillip Cartwright: *Developing Observation Skills*, McGraw-Hill, New York, 1974. (This is a very practical and lucid introduction to classroom observing. It emphasizes the importance of observations as the bases for instructional decision making. Numerous nontechnical suggestions are also given for recording classroom data.)

Good, Thomas L., and Jere E. Brophy: *Looking in Classrooms*, Harper and Row, New York, 1973. (An excellent book to aid in the development of better awareness of classroom behaviors. It deals with teachers' attitudes, classroom management, questioning strategies, motivation, and interaction-analysis instruments. Numerous research studies are surveyed, but the material is extremely readable.)

Hough, John B., and James K. Duncan: *Teaching: Description and Analysis*, Addison-Wesley, Reading, Mass., 1970. (An advanced, thorough, in-depth treatment of teaching as a rational, professional, and humane activity. The process of teaching is analyzed and described in four phases: curriculum planning, with emphasis on instructional objectives; instruction, with emphasis on interactions and communication; measurement; and evaluation.)

Ober, Richard L., Ernest L. Bentley, and Edith Miller: *Systematic Observation of Teaching*, Prentice-Hall, Englewood Cliffs, N.J., 1971. (Provides general guidelines for developing greater awareness and control of classroom behaviors as results of systematic observations. Two systems are described in detail—the "Reciprocal Category System" and "Equivalent Talk Categories." Much emphasis is placed on building and implementing instructional strategies.)

"Systematic Observation," *Journal of Research and Development in Education*, College of Education, University of Georgia, Athens, Ga., vol. 4, no. 1, Fall 1970. (Five observation instruments are presented along with background on the entire systematic observation movement. Research findings on the use of the systems are also included. The material is rather technical but very valuable if you wish to pursue the topic of systematic observations in greater depth.)

CHAPTER 9

ANALYZING AND INTERPRETING ELEMENTARY SCIENCE CURRICULA

243

If you were able to observe science being presented to children in many classrooms throughout the country, you would probably be amazed by the great variety of science programs being used or at least available for use. There's a myriad of resources to help you do science with children, and each year the number increases.

As was mentioned in Chapter 1, the major impetus for curriculum development in elementary science occurred in the 1960s, as one of the many reactions to Sputnik. It was widely believed that "national defense" and "science education" were linked, that we could no longer risk producing inferior students of science at any educational level; and that we had to upgrade both the quality and quantity of science education. Consequently, numerous science curriculum development projects were begun at the local, state, and national level; much of the funding came from the National Science Foundation. By the 1970s, many of these projects had failed; others were simply short-lived. But from these activities of the 1960s, elementary science education gained a new foothold among the more established subject areas.

Although several programs managed to survive both the competition of the 1960s and the criticism they received in the 1970s, three have emerged as the dominant products of curriculum development: *Science—A Process Approach* (S—APA, revised as S—APA II); *Science Curriculum Improvement Study* (SCIS, now available as Rand McNally SCIIS and SCIS II by American Science and Engineering); *Elementary Science Study* (ESS). (Some other "alphabet curricula," such as COPES, MINNEMAST, and USMES, are listed in the Suggested Readings at the end of this chapter.) Even though S—APA, SCIS, and ESS have not been as widely accepted and used as their developers and supporters had hoped, they have had an enormous influence on current, more conventional elementary science textbook programs. Therefore, it's important that you become familiar with these three programs in order to analyze and interpret elementary science curricula in general. As you proceed through this chapter, try to formulate answers to the following four important questions:

In what ways are S—APA, SCIS, and ESS similar?
What are their distinctive characteristics?
What are some examples of the strategies they use?
What's involved in evaluating and selecting elementary science curricula?

It's impossible to predict whether you will actually have the opportunity to use any of these programs with your students; but knowledge of these programs is important in itself. They provide excellent models for science with children and an enormous reservoir of ideas. Still, a caution is necessary before you proceed: There's probably no perfect elementary science program in existence. The main intent of this chapter is to help you become more skillful at analyzing and interpreting elementary science curricula that you can eventually use with your students. Try to consciously, honestly analyze and interpret what sorts of science will work best for you and your students.

Similarities among S–APA, SCIS, and ESS

Science—A Process Approach (S—APA), *Science Curriculum Improvement Study* (SCIS), and *Elementary Science Study* (ESS) all provide children with opportunities to learn by actively *doing* science rather than by passively watching or listening to or reading or memorizing science. Their approaches are inductive rather than deductive. Learning is elicited from the students rather than being imposed by the teachers. Children are encouraged to touch and handle things rather than admonished to keep hands off. The teacher is a questioner rather than an answerer, a receptive listener rather than a didactic lecturer, a suggester rather than a commander. The teacher is released from the role of all-knowing dispensor of facts and freed to say unashamedly, "I wonder about that, too; let's find out together." The emphasis is on discovering experiences cooperatively rather than covering content individually.

Many people (in fact, several hundreds) were involved in the development, testing, and subsequent revisions of each program. In the past, elementary science textbook programs were usually written by just a small team of authors—a program might have as few as two, and probably not more than eight, authors and a few consultants. The material generally went from the writers' desks to the editors', then to production, and then to the open market. Input from teachers and students was nonexistent or only slight. The development of the "alphabet curricula" was quite different. Teams of scientists, science educators, science supervisors, administrators, psychologists, and elementary teachers met together (usually in the summer) to brainstorm and develop ideas and strategies. During the following school year, hundreds of teachers tested these strategies with thousands of children in a variety of settings throughout the country. Their reactions and results were then analyzed so that the activities or units could be improved before they were ever placed on the market. No materials reached a new teacher's desk that had not been tried, tested, revised, and proven successful by other teachers and children.

Many of the topics of S—APA, SCIS, and ESS were the creative ideas of classroom teachers and were, therefore, very appropriate to the actual interests and intellectual levels of children. Since few if any printed materials were given to the children, science learning was no longer dependent on reading ability. The topics were also far more multidisciplinary than those from any previous programs. Science was meshed with mathematics, social studies, language development, and sometimes even art and music. These programs still maintained a balance among biological, physical, and earth sciences, but their approach to learning was much more holistic than that of previous science programs.

The developers of the three programs strongly believed in the importance of learning by *experience;* they believed that children's learning is enhanced by opportunities to interact with materials. Accordingly, packaged, ready-for-use kits of things to be manipulated were integral components of their programs. This feature had both positive and negative effects. The S—APA, SCIS, and ESS kits were greeted enthusiastically by students, but they caused some problems for administrators and teachers. In schools where science had been offered only through books, lectures, or demonstrations, the kits represented a new and seemingly exorbitant expense. The kits also created management problems. Where should the materials be stored? Who should be responsible for inventories, distribution, and maintenance? In an attempt to keep the prices as low as possible, companies sometimes sent inferior equipment. Sometimes a piece of equipment, or a specific setup, was so unfamiliar that teachers didn't know how to assemble it; and once it was assembled, they might not know how to use it. Teachers who were already insecure or unable to improvise became even more frustrated. But those who managed to use the kits as means rather than ends found a new world of exciting opportunities for science with children.

The basic questions *what*, *how*, and *why* (discussed in Chapter 8) were also important to the developers of these "alphabet curricula." Since all three were developed at about the same time, to upgrade elementary science education, the answer to the question *why* was quite similar for each. The major differences among these programs can be found in their answers to the questions *what* and *how*. Analysis of each program's components will reveal *what;* analysis of each program's instructional strategy will reveal *how*.

Analysis of S–APA II

The primary goal of the developers of *Science—A Process Approach* was to give elementary students (kindergarten through grade 6) numerous opportunities to perform the same sorts of actions as scientists. Before S—APA, science in the elementary school had been presented mainly as a noun—that is, as a body of facts pertaining to the natural world. In the S—APA program (both the original and the revised version), the emphasis is placed on science as a verb. The major question was, "What do scientists do that children can also do?" Eventually, thirteen "answers" were presented—eight appropriate particularly for children in kindergarten through grade 3, and five for children in grades 4 through 6. These thirteen processes or thinking operations are as follows:

Basic skills for children in the primary grades

Observing
Using space-time relationships
Classifying
Measuring
Using numbers
Communicating
Inferring
Predicting

Integrated skills for children in the intermediate grades

Controlling variables
Interpreting data

Defining operationally
Formulating hypotheses
Experimenting

Components

You'll have a better idea of how these processes were incorporated and implemented after you analyze the major components of the S—APA II program: the modules, the planning charts, and the optional components.

Software and Hardware: The Modules   The total S—APA II program consists of 105 modules or units. *Each* module consists of a teacher's instruction booklet (seven to fifteen pages); an accompanying kit of equipment and materials; overhead transparencies; duplicating masters; and self-instruction booklets and data-collecting sheets sufficient for a class of thirty children. Modules can be used separately or as clusters related to certain skills or content areas (such as metrics, reading readiness, or environmental studies)—or as complete "mod pods" consisting of fifteen modules for each grade level.

Each *teacher's instruction booklet* contains the following:

Module number. This identification number indicates the module's placement within the total program sequence (the modules are numbered 1 to 105). For example, Module 13 is one of the later kindergarten units; Module 76 is the beginning module for grade 5.

Process title and sequence letter. Although one process skill often overlaps another, in each module emphasis is placed on a particular process. In Module 88, for example, the process title and sequence letter appear as "Defining Operationally/f." In other words, this is the sixth module dealing with the process of defining operationally.

Content title. Each module also has a very brief descriptive title of the content or subject matter that the children will encounter. (There is a wealth of content in S—APA II, even though the major emphasis is on the processes. It's impossible to perform the processes without the substance of subject matter.) Table 9-1 shows the complete listing of S—APA II module numbers, process titles with their sequence letters, and content titles.

Module objectives. These are handled in a unique manner. Although the modules have an average of three to four objectives each, only *nine* terms referring to performance—"performance terms," or "action words"—are used in the objectives throughout the 105 modules: *name, describe, order, demonstrate, identify, distinguish, construct, state a rule,* and *apply a rule.*

Sequence of module objectives. S—APA II is one of the few curricula that tell you very clearly where you are now (in terms of the total program), where you should have been before you got here (that is, the prerequisite experiences or modules), and the options for the next choices.

Rationale. This is an invaluable aid, especially for the neophyte. Depending on the particular module, it provides background information, some clarification on

Table 9-1
S–APA II MODULES

MODULE	PROCESS AND SEQUENCE	CONTENT TITLE
1	Observing a	Perception of Color
2	Space/Time a	Recognizing and Using Shapes
3	Observing b	Color, Shape, Texture, and Size
4	Classifying a	Leaves, Nuts, and Seashells
5	Observing c	Temperature
6	Space/Time b	Direction and Movement
7	Observing d	Perception of Taste
8	Measuring a	Length
9	Using Numbers a	Sets and Their Members
10	Space/Time c	Spacing Arrangements
11	Observing e	Listening to Whales
12	Space/Time d	Three-Dimensional Shapes
13	Using Numbers b	Numerals, Order, and Counting
14	Classifying b	Animals and Familiar Things
15	Observing f	Perception of Odors
16	Classifying c	Living and Nonliving Things/ Trees in Our Environment
17	Observing g	Change
18	Observing h	Using the Senses
19	Observing i	Soils
20	Using Numbers c	Counting Birds
21	Observing j	Weather
22	Communicating a	Same but Different
23	Measuring b	Comparing Volumes
24	Measuring c	Metric Lengths
25	Communicating b	Introduction to Graphing
26	Measuring d	Using a Balance
27	Communicating c	Pushes and Pulls
28	Observing k	Molds and Green Plants
29	Space/Time e	Shadows
30	Using Numbers d	Addition through Ninety-Nine
31	Communicating d	Life Cycles
32	Classifying d	A Terrarium
33	Inferring a	What's Inside?
34	Measuring e	About How Far?
35	Space/Time f	Symmetry
36	Observing f	Animal Responses
37	Measuring f	Forces
38	Predicting a	Using Graphs
39	Measuring g	Solids, Liquids, and Gases
40	Inferring b	How Certain Can You Be?
41	Measuring h	Temperature and Thermometers
42	Classifying e	Sorting Mixtures
43	Communicating e	A Plant Part That Grows
44	Predicting b	Surveying Opinion
45	Space/Time g	Lines, Curves and Surfaces

Row groups (left spanning labels): Modules 1–15 = KINDERGARTEN; Modules 16–30 = FIRST GRADE; Modules 31–45 = SECOND GRADE.

Table 9-1 (Continued)

MODULE	PROCESS AND SEQUENCE	CONTENT TITLE
46	Inferring c	Observations and Inferences
47	Communicating f	Scale Drawings
	Communicating g	A Tree Diary
48	Predicting c	The Bouncing Ball
49	Measuring i	Drop by Drop
50	Predicitng d	The Clean-Up Campaign
51	Space/Time h	Rate of Change
52	Inferring d	Plants Transpire
53	Predicting e	The Suffocating Candle
54	Measuring i	Static and Moving Objects
55	Observing m	Sprouting Seeds
	Observing n	Magnetic Poles
56	Classifying f	Punch Cards
57	Communicating h	Position and Shape
58	Inferring e	Liquids and Tissue
59	Using Numbers e	Metersticks, Money, and Decimals
60	Space/Time i	Relative Motion
61	Inferring f	Circuit Boards
62	Controlling Variables a	Climbing Liquids
63	Interpreting Data a	Maze Behavior
64	Defining Operationally b	Cells, Lamps, Switches
65	Interpreting Data b	Minerals in Rocks
66	Controlling Variables b	Learning and Forgetting
67	Interpreting Data c	Identifying Materials
68	Interpreting Data d	Field of Vision
69	Defining Operationally h	Magnification
70	Formulating Hypotheses a	Conductors and Nonconductors
71	Controlling Variables c	Soap and Seeds
72	Controlling Variables d	Heart Rate
73	Formulating Hypotheses b	Solutions
74	Defining Operationally c	Biotic Communities
75	Interpreting Data e	Decimals, Graphs, and Pendulums
76	Interpreting Data f	Limited Earth
77	Controlling Variables e	Chemical Reactions
78	Formulating Hypotheses c	Levers
79	Formulating Hypotheses d	Animal Behavior
80	Defining Operationally d	Inertia and Mass
81	Defining Operationally e	Analysis of Mixtures
82	Controlling Variables f	Force and Acceleration
83	Formulating Hypotheses e	Chances Are
84	Interpreting Data g	Angles
85	Interpreting Data h	Contour Maps
86	Interpreting Data i	Earth's Magnetism
87	Interpreting Data j	Wheel Speeds
88	Defining Operationally f	Environmental Protection
89	Defining Operationally g	Plant Parts
90	Interpreting Data k	Streams and Slopes

THIRD GRADE (modules 46–60)
FOURTH GRADE (modules 61–75)
FIFTH GRADE (modules 76–90)

Continued on page 250.

Table 9-1 (Continued)

MODULE		PROCESS AND SEQUENCE	CONTENT TITLE
	91	Defining Operationally h	Flowers
	92	Formulating Hypotheses f	Three Gases
	93	Defining Operationally i	Temperature and Heat
	94	Controlling Variables g	Small Water Animals
	95	Interpreting Data 1	Mars Photos
SIXTH GRADE	96	Experimenting a	Pressure and Volume
	97	Experimenting b	Optical Illusions
	98	Experimenting c	Eye Power
	99	Experimenting d	Fermentation
	100	Experimenting e	Plant Nutrition
	101	Experimenting f	Mental Blocks
	102	Experimenting g	Plants in Light
	103	Experimenting h	Density
	104	Experimenting i	Viscosity
	105	Experimenting j	Membranes

the process to be used, suggestions for instructional strategies, cautions, advice for advance preparation, and—in general—comments on why the module is worth presenting.

Vocabulary list. This consists of the names or the scientific terms for the new concepts the children will encounter in the module. These terms should not be introduced before the activities; they should not be "drilled" into the students. They should be presented only after the children have experienced their meaning. The S—APA *Guide for Inservice Instruction* presents one of the best pedagogical suggestions for handling any vocabulary work with children—the idea that "Adam named the animals *after* he saw them," not before (1973, p. 106).

Introduction to the module. Usually this is the suggested motivational technique that the teacher can use to stimulate the students' interest—posing a problem to be solved, for example, or explaining the importance of the process to be encountered. It is the first component of the actual instructional procedures suggested in the teacher's instruction booklet. All the preceding components (that is, the coding systems of numbers and letters, the titles of process and content, the objectives, the sequence, the rationale, and the vocabulary list) serve mainly as preliminary considerations.

Module activities. These constitute the meat of the module. Each unit is divided into several activities. Some activities can be presented in one science period; others may extend over several lessons. It all depends on the children's abilities, the nature of the activity, and, of course, the decision of the teacher. The activities are numbered, and generally arranged sequentially in order of difficulty. Sometimes, however, one activity may serve only as a reinforcement for another. If the children have already mastered the corresponding objective, that activity can be omitted. Often these sorts of activities appear under the heading "Optional Activity." Each activity presents very specific suggestions for the instructional procedures a teacher might use—directions, questions to ask the children (even possible responses to anticipate), diagrams, charts, and so on.

Generalizing experience. This is the culminating experience, the capstone, the opportunity to apply what was previously learned to a new situation.

Evaluation components. Each S—APA II instruction booklet provides the teacher with *two* evaluation techniques. One is the *group competency measure* (also referred to as the *appraisal* in the primary modules), which can be administered to the entire class. The other is the *individual competency measure*, which can be administered when necessary to individual students. The items in both of these instruments correspond directly to the objectives of the module. Therefore they require the children, or individual children, to actually demonstrate one or more of the nine actions specified in the program goals. They provide the teacher with immediate diagnostic feedback concerning the degree of learning acquired. Such information is invaluable for the next instructional decision—whether to provide remediation or enrichment, or to move on to the next module.

Materials list. At the end of each introduction, each specific activity, each generalizing experience, and each group and individual competency measure, a list of materials is presented. If the materials are supplied in a particular module, the module number is indicated in parentheses. If they are to be provided by the teacher or the children, no module number is cited.

Program Organizer: The Planning Charts   The entire sequence of S—APA II has been presented on two charts: the "Basic Processes Chart," which shows Modules 1 through 60; and the "Integrated Skills Chart," which shows Modules 61 through 105.

The planning charts provide the following information for each of the 105 modules: (1) module number, (2) process and sequence letter, (3) content title, (4) summary of major objectives, and (5) "premod" numbers—that is, the numbers of prerequisite modules. Generally, the "premods" are based on mastery of a *process* rather than of content.

The initial portion of the planning chart for kindergarten is shown in Figure 9-1. Each vertical column of the planning chart represents a level or stage of difficulty. Modules within the same column have the same degree of difficulty and therefore may be selected in any order. Modules have also been organized in terms of "clusters" (indicated by the shading). These clusters represent similar emphases in terms of process or content. Figure 9-1 shows that the kindergarten teacher may begin with Module 1, 2, 6, or 7; or with a cluster consisting of Modules 1, 2, 3, and 4, then Modules 5, 6, 8, and 12, then Modules 7, 11, 14, and 15. Ideally, a teacher should strive to have the children successfully complete the modules in the lowest-numbered column before proceeding to the next higher-numbered column, or the clusters with the lowest-numbered modules before the clusters with the next higher-numbered modules. Table 9-2 summarizes the planning charts for the entire program.

The planning charts provide much valuable information to the teacher, which can make the decision-making process easier. A teacher can identify at a

I	II	III	IV

MODULE 1
Observing/a
PERCEPTION OF COLOR
Identifying and naming the primary and secondary colors. Identifying other colors as being similar to the primary colors.

MODULE 2
Using Space/Time/a
RECOGNIZING AND USING SHAPES
Identifying and naming common two-dimensional shapes. Recognizing common shapes in the environment.

MODULE 3
Observing/b
COLOR, SHAPE, TEXTURE, AND SIZE
Identifying and naming objects and construction groupings on the basis of color, shape, texture, and size.
PREMODS: 1, 2

MODULE 4
Classifying/a
LEAVES, NUTS, AND SEASHELLS
Describing a method of classifying objects. Constructing a system of classification according to variations in a single characteristic.
PREMOD: 3

MODULE 10
Using Space/Time/c
SPACING ARRANGEMENTS
Constructing, naming and identifying two-dimensional shapes formed by arrangements of objects.
PREMOD: 2

MODULE 9
Using Numbers/a
SETS AND THEIR MEMBERS
Identifying sets of objects when given the characteristics of their members. Identifying the set that has the fewest members, the most members and equivalent members.

MODULE 13
Using Numbers/b
NUMERALS, ORDER, AND COUNTING
Identifying and naming numerals and ordinal names from 1 to 5. Ordering sets that have 0-12 members.
PREMODS: 3, 9

MODULE 5
Observing/c
TEMPERATURE
Distinguishing between temperatures using sensory input, and between different places and times using a color-coded thermometer.
PREMOD: 1

MODULE 6
Using Space/Time/b
DIRECTION AND MOVEMENT
Naming and demonstrating movements or directions using the words up, down, forward, backward, right, and left. Distinguishing between objects that are moving and those that are not moving.

MODULE 8
Measuring/a
LENGTH
Ordering objects according to length. Demonstrating the length of two objects by comparing them to a third.

MODULE 12
Using Space/Time/d
THREE-DIMENSIONAL SHAPES
Identifying and naming two-dimensional shapes as components of three-dimensional shapes. Identifying and naming spheres, cubes, cylinders, pyramids, and cones.
PREMOD: 2

MODULE 14
Classifying/b
ANIMALS AND FAMILIAR THINGS
Constructing a classification of a set of objects into two or more groups, based on the way in which the objects can be used. Constructing another classification of the same objects, and describing the characteristics used to classify the objects.
PREMODS: 4, 7, 11

MODULE 7
Observing/d
PERCEPTION OF TASTE
Identifying tastes that are sweet, sour or salty. Distinguishing between food tastes as being similar to or different from each other.

MODULE 11
Observing/e
LISTENING TO WHALES
Naming and identifying a sound as being longer or shorter, louder or softer, higher or lower, or more like one sound than another.

MODULE 15
Observing/f
PERCEPTION OF ODORS
Distinguishing between objects that have an odor and those that do not. Identifying familiar odors as being similar to or different from others.

Figure 9-1 Kindergarten portion of the S-APA II Basic Process Chart. (From *Science—A Process Approach*, II, 1974. Used by permission of the American Association for the Advancement of Science.)

Table 9-2
S–APA PLANNING CHARTS: SUMMARY

GRADE	MODULES	COLUMN STAGES
Kindergarten	1–15	I–IV
First	16–30	V–VIII
Second	31–45	IX–XII
Third	46–60	XIII–XVII
Fourth	61–75	XVIII–XXI
Fifth	76–90	XXII–XXVI
Sixth	91–105	XXVII–XXXIII

glance the levels of difficulty as well as the clusters of related processes or content. (A system of color coding makes the clusters more obvious on the actual planning charts. Modules dealing with the physical sciences are color-coded in red or orange. Green and brown clusters indicate modules which emphasize the life sciences.) The planning charts also show the "premods," or prerequisites for specific modules, and indicate which modules require long-term data collecting and which modules are most important for new students in the intermediate grades who have not experienced the S—APA II program before.

Some Extras: The Optional Components   The 105 K–6 modules with their teachers' instruction booklets, the kits of materials, the "Basic Processes Chart" for Modules 1 through 60, and the "Integrated Skills Chart" for Modules 61 through 105 are the most important components of the S—APA II program. But the following optional components are also available and quite helpful:

Commentary for teachers. This is an excellent resource book which contains a thorough description of the S—APA program, several self-instructional activities on each of the thirteen processes, and seventeen background papers on topics such as measuring temperature and heat, velocity and acceleration, the cell, density, magnetism, and photosynthesis.

Orientation package. This kit contains five filmstrips and cassettes for in-service training along with a limited supply of materials similar to those that the children will be using. For example, chemically treated masters for duplicating worksheets and "latent image markers" are included. Copies from these masters contain invisible answers which will appear only when the correct response is rubbed with the special "latent image marker."

Competency tracking cards. These cards, based on the keysort or punchcard model, can be used as a recordkeeping device on each student's attainment of the objectives of each module. One card covers the basic-skill modules (1 through 60); another covers the integrated-skill modules (61 through 105). If the tracking cards are a part of the children's cumulative folders, each succeeding teacher can readily note each child's placement in and progress through the program.

Conversion kits. These are packages of the S—APA II instruction booklets along with materials not present in the original S—APA program. As their name indicates, they aid in converting from the original S—APA to S—APA II.

Differences between S–APA and S–APA II

Since occasional references have been made to the original program (S—APA) and S—APA II, some comments on the differences between the two might be helpful. Even if you never need to compare them, it's interesting to note how a curriculum evolves.

As is true with most revisions, both major and minor changes were made. The following represent the major modifications in S—APA II:

1. *Much greater flexibility in the hierarchy or sequence of the modules is available.* S—APA II is still a highly sequential program; prerequisites are still very important. But a teacher has far greater latitude in designing and selecting the sequence of modules most appropriate for specific children than the original program allowed for.
2. *There is less emphasis on mathematics and more emphasis on reading.* (Mathematical concepts and skills are still an integral part, however.) S—APA II includes forty-six self-instruction booklets of printed and pictorial materials for individualized learning. In the original program, the printed materials for use by students consisted of worksheets or recordkeeping sheets only.
3. *S—APA II is more responsive to the needs and interests of the times.* It includes several new modules dealing with environmental and ecological topics.

Because of the new emphases, several minor changes are also evident. There is less material to cover in a school year, since some activities have been omitted or combined with others. The addition of some new units and new instructional procedures necessitated some new hardware and software. The new suggested organization of fifteen modules per grade level has resulted in a new numbering and lettering system for S—APA II.

Additional Considerations on S–APA II

Both the original version of S—APA and S—APA II emphasize a general hierarchical approach to instruction rather than a specific or unique instructional strategy for teachers. Therefore it is difficult to present a single sample activity for S—APA as is done for the SCIS and ESS programs on pages 259 and 269. Instead, you might refer back to Table 9-1, select a specific process, and track its sequence through the entire program. Or you might reexamine several of the activities described in Part Two to get the flavor of a S—APA module. For example, Module 25, "Introduction to Graphing," is summarized on page 141. Module 48, "The Bouncing Ball," was the springboard for "Bouncing Ball

Baffles," pages 160 to 162; Module 53, "The Suffocating Candle," provided the framework for "Burned Out," pages 162 to 164; and Module 78, "Levers," was the basis for the "A Balanced Relationship," pages 166 to 169.

As is true with all curriculum programs, there are some pros and cons to *Science—A Process Approach.* But you'll be in a better position to make judgments after you have analyzed the other programs.

Analysis of SCIS

The primary goal of *Science Curriculum Improvement Study* is to aid in the development of scientific literacy. Scientifically literate citizens should be able to identify the methods of inquiry or the processes of science as well as the products of inquiry or the conceptual structure of science. (A scientifically literate teacher would be able to do science with children, as the title of this book imperatively and declaratively suggests.) In order to achieve the goal of scientific literacy, the SCIS program emphasizes the laboratory or experiential approach to learning the content of science. The secondary goal is to help children develop more logical thinking skills. Two general approaches are used in SCIS to achieve this secondary goal and the primary goal: a sequence or hierarchy of concepts, and a distinct instructional strategy. Each of these approaches will be explained in detail, but some initial clarification is needed concerning the development and production of the program.

Development of the original SCIS program began in the 1960s at the University of California at Berkeley with funds from the National Science Foundation. Rand McNally became the commercial publisher and marketing agent for this program in 1969. Two additional programs based on the original SCIS were marketed in 1978: Rand McNally SCIIS, and SCIS II by American Science and Engineering. (Some other programs have also incorporated much of the original SCIS material.) These three programs—the original SCIS published by Rand McNally, the new Rand McNally SCIIS, and SCIS II by American Science and Engineering—present slightly different concepts, but in similar sequences. Table 9-3 shows the similarities and differences.

SCIS Concepts

Each version of the SCIS program offers a hierarchy of concepts, so that what is presented in an earlier grade level is expanded in succeeding grades. A two-unit sequence is used for all grades except kindergarten. Half of each year's curriculum deals with life science and the other half with physical science. (The Rand McNally SCIIS program includes earth science throughout the entire program.) A teacher has the option of dealing with one unit in its entirety before presenting the other or of interspersing both throughout the school year. It should be noted that the life science–earth science sequence took a strongly ecological approach before environmental or ecological concerns became

Table 9-3
COMPARISONS OF THE ORIGINAL SCIS AND PROGRAMS BASED UPON SCIS

	GRADE LEVEL	UNIT	CONCEPTS	(RAND) SCIS	(RAND) SCIIS	(AS AND E) SCIS II
	K	Beginnings	Color	X	X	Early Childhood Curriculum—A Piaget Program
			Shape	X	X	
			Texture	X	X	
			Odor	X	X	
			Sound	X	X	
			Size	X	X	
			Quantity	X	X	
			Position	X	X	
			Organisms	X	X	
LIFE/EARTH SCIENCE	1	Organisms	Organism	X	X	X
			Birth	X	X	X
			Death	X	X	X
			Habitat	X	X	X
			Food Web	X		X
			Detritus	X		X
			Food Chain		X	
			Decay		X	
			Growth			X
	2	Life Cycles	Growth	X	X	X
			Development	X	X	X
			Life Cycle	X	X	X
			Genetic Identity	X	X	X
			Biotic Potential	X		X
			Generation	X		X
			Plant and Animal	X	X	X
			Metamorphosis	X	X	X
			Germination			X
	3	Populations	Population	X	X	X
			Food Chain	X		X
			Food Web	X	X	X
			Community	X		X
			Plant Eater		X	X
			Biotic Potential		X	
			Animal Eater		X	X
			Plant-Animal Eater		X	X
			Predator-Prey		X	X
			Dispersal			X
	4	Environments	Environment	X	X	X
			Environmental Factor	X	X	X
			Range	X	X	X
			Optimum Range	X		X
			Biotic		X	
			Abiotic		X	

Table 9-3 (Continued)

	GRADE LEVEL	UNIT	CONCEPTS	(RAND) SCIS	(RAND) SCIIS	(AS AND E) SCIS II
LIFE/EARTH SCIENCE	4	Environments (continued)	Seasonal Change			X
			Temperature			X
			Response			X
			Optimum		X	
	5	Communities	Photosynthesis	X	X	X
			Community	X	X	X
			Food Transfer	X	X	X
			Producers	X	X	X
			Consumers	X	X	X
			Decomposers	X	X	X
			Raw Materials	X	X	X
			Pyramid of Numbers		X	
			Reproduction		X	X
			Competitors		X	
			Food Source			X
			Food Cycle			X
	6	Ecosystems	Ecosystem	X	X	X
			Water Cycle	X	X	X
			Food-Mineral Cycle	X	X	X
			Oxygen–Carbon Dioxide Cycle	X	X	X
			Pollutant	X		X
			Pollution		X	
			Evaporation			X
			Condensation			X
			Gas			X
PHYSICAL/EARTH SCIENCE	1	Material Objects	Object	X	X	X
			Property	X	X	X
			Material	X	X	X
			Serial Ordering	X	X	X
			Change	X		X
			Evidence	X	X	X
	2	Interaction and Systems	Interaction	X	X	X
			Evidence of Interaction	X	X	X
			System	X	X	X
			Interaction-at-a-distance	X	X	X
			Electric Circuit			X
			Magnetic Interaction			X
	3	Subsystems and Variables	Subsystem	X	X	X
			Evaporation	X	X	X
			Histogram	X	X	X
			Solution	X	X	X
			Variable	X	X	X
			Temperature			X

Continued on page 258.

Table 9-3 (Continued)

	GRADE LEVEL	UNIT	CONCEPTS	(RAND) SCIS	(RAND) SCIIS	(AS AND E) SCIS II
PHYSICAL/EARTH SCIENCE	4	Relative Position and Motion	Reference Object	X	X	
			Relative Position	X	X	
			Relative Motion	X	X	
			Polar Coordinates	X	X	
			Rectangular Coordinates	X	X	
	4	Measurement, Motion, and Change	Reference Frame			X
			Relative Position			X
			Polar Coordinates			X
			Rectangular Coordinates			X
			Relative Motion			X
			Reference Object			X
			Change			X
			Distance			X
			Direction			X
			Measurement			X
	5	Energy Sources	Energy Transfer	X	X	X
			Energy Chain	X	X	X
			Energy Source	X	X	X
			Energy Receiver	X	X	X
			Temperature Change			X
	6	Scientific Theories	Scientific Theory		X	
			Magnetic Field		X	
			Electricity		X	
			Light Ray		X	
	6	Models: Electric and Magnetic Interactions	Scientific Model	X		
			Magnetic Field	X		
			Electricity	X		
	6	Modeling Systems	Model			X
			Electricity			X
			Magnetism			X
			Circuit			X
			Electrical Energy			X
			Air Temperature			X
			Barometric Pressure			X
			Atmosphere			X

popularized. Although the titles of the physical science units may appear somewhat unconventional, they become much more meaningful as their related concepts are experienced.

The selection of concepts was highly influenced by Piagetian considerations. A deliberate attempt was made to select beginning concepts appropriate for the preoperational child. (Examine the concepts presented in the kindergarten unit "Beginnings," shown in Table 9-3.) These are followed by concepts geared to the concrete-operational child, and finally by concepts and related skills which will facilitate the development of more formal reasoning. For example, the opportunities for children to operate on the regularities of the physical world are increased throughout the program. They observe patterns of occurrences and gradually discover that certain events are predictable. Additional practice is provided for making and testing general statements of relationships or hypotheses. Children develop plans for controlling variables by testing only one factor at a time. With guidance from the teacher, children are encouraged to use the logic of implication by developing and testing statements of the form "If this, then that." These are some of the ways in which the sequence of concepts helps children to move from concrete to more logical thinking. The learning cycle used throughout the program is also particularly conducive to concept formation.

SCIS Instructional Strategy

The SCIS instructional strategy is one of the most distinctive features of the program. It is a three-stage strategy that can be used with any science program as well as with many other subject areas. It's also a fairly easy plan to implement.

Suppose that a fifth-grade teacher wanted the children to learn the names of the parts of seeds, such as *cotyledon* and *embryo*. The following situations show how the SCIS three-stage strategy would be used: each represents one stage of the process.

Situation 1   Each child is given some presoaked lima beans and navy beans, and a hand lens. The children are then encouraged to observe the seeds and investigate their characteristics. The teacher might ask any of the following questions:

"What can you tell me about these seeds on the basis of your observations?"

"How are they alike? How are they different?"

"How might these seeds be described to someone who had never experienced them before?"

The children are given time to investigate, interact with the seeds, examine the parts found, and share their observations—they are even given time to wonder. This is the *exploration stage*. The emphasis is on the children's explorations. The teacher is there to guide, encourage, suggest, and—most important—listen to and watch the interactions.

Figure 9-2 Seed parts for an invention stage.

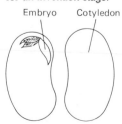

Embryo   Cotyledon

Situation 2   The children are then asked to share their observations so that the teacher can determine if the two seed parts were actually noted. A wall chart or large drawing such as Figure 9-2 could then be displayed. The children are asked to observe or handle either one of the large, meaty halves of the seed as the same part is identified on the drawing by the teacher. "This part of the seed has a special name," the teacher explains. "It's called the *cotyledon*." This is the *invention stage* of the SCIS strategy. It is extremely unlikely, of course, that a child would accidentally decide that *cotyledon* would be a good name for that particular seed part. The teacher, therefore, simply presents the name that was originally "invented" by some scientist. The invention stage is never an end in itself. It must always be the framework for the next stage.

Situation 3 New presoaked seeds are given to the children—such as kidney beans, pinto beans, mung beans, or peas. The children are then asked to identify and describe the cotyledons and embryos of these seeds and compare them with those presented during the invention stage. That is, they are given the opportunity to apply in a new situation concepts that were originally explored by them and subsequently named for them. This opportunity to apply their new understanding is the *discovery stage*. Activities in the discovery stage may be suggested by the teacher or the students. The purpose of the discovery stage can be to reinforce, refine, or expand the previous stages. The greater the *variety* of discovery activities (rather than the number of repetitions), the less chance there is that the children will form misconceptions or stereotypes. The discovery stage can also stimulate ideas for new exploration activities.

Each component of the SCIS three-stage strategy can be used separately; the stages need not be combined in a single lesson. For instance, at the end of the discovery stage in the example described, students might inquire about the purpose of the cotyledon. The exploration stage could then center on "How can we find out?"—with the children exploring various possibilities and perhaps planting seeds. This exploration stage could then be extended over a longer period of time. The invention stage might deal with the term *germination*, rates of germination, or a new name for the cotyledon—*seed leaves*. The discovery stage could involve changing the planting conditions, determining the germination rates of other seeds, or even planting seeds with one or both portions of the cotyledon removed.

This strategy can be applied in a multitude of situations, but care must be exercised in how it is used. *First,* none of the stages can stand alone indefinitely; they are interrelated and interdependent. Children need ample time to explore and discover; but they also need some guidance. Merely providing things to manipulate and saying "Here they are; go to it!" can lead to chaos. *Second,* the invention stage must relate to what the children have been exploring—not to the teacher's own interests or desire to expound facts. *Third,* both the number of concepts and the distracting information should be limited.

I recommend adding a step to the basic three-stage SCIS strategy; invite children to formulate their own operational definitions *before* the teacher's presentation of the terms previously invented by scientists. Questions having to do with the invention stage can be directed to what the children have been doing and observing. This will help the children relate their own concrete, personal experience to the new terms. The teacher should also provide operational definitions during the invention stage rather than relying only on formal definitions.

The skills which seem most necessary for the effective use of the SCIS instructional strategy are the ones familiar from Chapter 8—namely, observing, questioning, and listening. These same skills are needed in implementing SCIS

II, in which a comparable strategy is called *exploration, concept lessons*, and *application*.

Program Components

Because of the variety of SCIS programs available, only general topics common to all the programs will be described here. The following are the most important components of the program:

Teacher's guides. These include objectives for major parts of each unit, background information, photographs and illustrations on techniques, procedures for implementing the activities and applying the instructional strategy, clues for advance preparation, suggestions for questions, summaries, and evaluation techniques.

Student manuals. There are student manuals for all units except the kindergarten unit, "Beginnings." These are data-recording booklets rather than mini-textbooks. They also contain ideas for extension activities and can serve as additional means for evaluating children's performance.

Student activity cards. These too exist for all units except "Beginnings." They can be used in several ways for a variety of purposes. Activity cards may be selected by an individual child, the teacher, a group of children, or the entire class. The cards may serve as enrichment, remediation, a way to help a transfer student adjust to the program, and a means of interrelating the science topics with other subject areas. (These types of activity cards do not accompany the original SCIS program.)

Kits. Kits of things to be manipulated and printed charts or illustrations accompany each unit. (The measuring devices in the newer SCIS programs use only metrics.) Some of the items supplied are unique; others are usually available at stores. About 80 to 90 percent of the physical science materials are supplied for about thirty children in a class. The live organisms must be ordered separately from the publishing company or from other sources. The organisms for the life science half of the programs have been selected on the basis of appeal to children, ease in handling and maintenance, hardiness, and the insignificance of their effect on the environment (that is, their absence from their usual environment and their presence in a classroom will not have detrimental effects). Some of the more familiar organisms are guppies, crickets, chameleons, mealworms, and various aquatic plants. Refill packages and conversion packages are also available.

Evaluation packets. These are an optional component for the units, since evaluation techniques are incorporated in the teacher's guide.

Concept charts. These give a brief overview of the entire K–6 program along with the concept sequence used.

Science Activities for the Visually Impaired (SAVI) is not exactly a component of the SCIS program, but it is an interesting adaptation of some of its ideas for special children. SAVI also demonstrates the dynamic nature of elementary science curriculum development by providing interesting and exciting science encounters for all children.

Analysis of ESS

If you could choose only one word to describe the emphasis of each of these three major curricula, the most likely choices would be *process* for Science–A Process Approach, *content* for *Science Curriculum Improvement Study*, and *experience* for *Elementary Science Study*. Providing children with opportunities to experience the phenomena of science and to derive their own unique discoveries is the major purpose of ESS.

The original developers of ESS took a different view from the developers of S—APA and SCIS. They chose not to develop a specific scope and sequence, for several reasons. *First*, they believed that there was no way of knowing what children in the final third of the twentieth century will need for their adulthood in the first half of the twenty-first century. *Second*, they contended that no one really knows the best way to sequence instruction or learning encounters; the empirical data are insufficient and narrow. *Third*, they thought that no curriculum developers can devise a program to meet the needs of all possible constituents; the content and sequence of any program should be based on the unique needs of each school. *Fourth*, they believed that self-directed learning is the best form of learning for children as well as for teachers, and that therefore teachers should also be given options. Consequently, ESS was designed not as an elementary science program in the usual sense but as a collection of fifty-six interdisciplinary units spanning various grade levels. Teachers at a specific school (or in a specific school system) can select what they consider the most appropriate combinations of units, or they can choose individual units to supplement existing programs.

ESS Units and Related Components

Each unit of ESS had to meet certain criteria during the developmental and testing stages before it became available commercially: (1) It had to actively engage children in interacting with things rather than ideas. (2) The things encountered had to cause children to ask a wide variety of questions. (3) The things had to elicit not only actions from each child but also interactions among children encountering similar phenomena. Of the fifty-six units which met these criteria, twenty-one are related to physical science, four to earth science, nineteen to biology topics, and twelve to general skills such as perceptual development, manipulative skills, mapping, graphing, measuring, classifying, diagraming, and thinking skills in general. (See Table 9-4.)

The grade spans were determined by the results of trial testing. You may eventually discover, however, that the applicability of certain units can vary according to the interests and skills of your students. The units cover astronomy,

Table 9-4 (Continued)

GENERAL SKILLS UNITS	GRADE									
	K	1	2	3	4	5	6	7	8	9
Match and Measure	x	x	x	x						
Primary Balancing	x	x	x	x	x					
Pattern Blocks	x	x	x	x	x	x	x			
Geo Blocks	x	x	x	x	x	x	x			
Tangrams	x	x	x	x	x	x	x	x	x	
Musical Instrument Recipe Book	x	x	x	x	x	x	x	x	x	x
Attribute Games and Problems	x	x	x	x	x	x	x	x	x	x
Printing		x	x	x	x	x	x			
Structures			x	x	x	x	x			
Whistles and Strings					x	x				
Peas and Particles					x	x	x			
Mapping						x	x	x		

The amount of time needed to complete a unit varies greatly, depending on the particular topic selected, the interest of the children, and the decision of the teacher. (Flexibility is also a key descriptor of the ESS units.) Suggested times are given in each teacher's guide, along with an abundance of additional information, such as:

Accounts of previous teachers' efforts, which are descriptive rather than prescriptive.

A suggested sequence of activities and possible questions to be pursued.

Background information, suggestions for advance preparation when necessary, descriptions of materials to be used by the children.

Examples of students' discoveries such as drawings, charts, graphs, reports.

The revisions which were begun in 1974 include additional suggestions for planning, management, and recordkeeping. The revised units also indicate possible ways of integrating art, language development, music, and social studies with each specific unit topic as well as with other ESS units. The revision which teachers seem to have welcomed the most is the addition of a brief description of the skills and content to be encountered in each unit; this makes evaluation more manageable.

McGraw-Hill Evaluation Program for ESS Too often in the past a teacher might comment at the completion of an ESS unit, "The students and I thoroughly enjoyed our experiences, but I'm not sure what was actually accomplished. How can I evaluate the results?" Similar comments were also voiced by school administrators and parents concerned about accountability or simply about children's learning. *The McGraw-Hill Evaluation Program for ESS* (Aho et al., 1974) was developed to provide a means for identifying the behaviors expected from students (that is, the objectives to be met) in each of the fifty-six units. These behaviors are related to the following evaluative categories:

I. Rational thinking processes—observation, classification, measurement, collection and organization of data, inference and prediction, identification and control of variables, making and testing of hypotheses, and process synthesis or experimentation.
II. Manipulation of materials and equipment.
III. Communication of data, reactions, and conclusions—orally, in writing, nonverbally, graphically.
IV. Attainment of concepts through recall or application.
V. Development of positive attitudes—curiosity, persistence, active participation, and increased self-confidence.

Because the units have varying grade-level applicability (see Table 9-4), the specific objectives are gradated according to three degrees of mastery: basic (or minimal achievement), intermediate, and advanced. The evaluation program can assist a teacher in systematically observing and evaluating students' progress. If the philosophy and original intentions of ESS are to be preserved, however, the evaluation program must remain a means rather than an end in itself, and therefore the teacher must be flexible.

Special Education Guide for ESS With two grants from the National Science Foundation, Daniel Ball conducted a three-year study to determine which ESS units were most beneficial for educable mentally retarded, trainable mentally retarded, emotionally disturbed, and learning-disabled children. More than fifty special education teachers participated in the project, using selected ESS units in classrooms from the primary grades through high school. The major results of the study are summarized in *ESS/Special Education Teacher's Guide* (Ball, 1978). This guide contains valuable suggestions for special education teachers who want to give their children experiences in science. For example, thirty-one ESS units are identified which have been proven beneficial for primary children (grades 1 to 3), intermediate students (grades 4 to 6), junior high school students (grades 7 to 9), and high school students (grades 10 to 12) who are either EMR (educable mentally retarded, that is, with IQs of 60 to 75), TMR (trainable mentally retarded, that is, with IQs of 35 to 59), ED (emotionally disturbed), or LD (learning-disabled). Table 9-5 summarizes the applicability of six units which improve perceptual development and seven which improve psychomotor development.

Ball's guide also specifies objectives for each of the thirty-one ESS units that could be suitable for special children. Suggestions are given for beginning the units, continuing them, and relating the science encounters to objects or events in the daily lives of the children. Observation and evaluation checklists consisting of the unit objectives are available for reproduction if desired. Specific research

Table 9-5
SPECIAL EDUCATION UNITS IN ESS

UNITS FOR PERCEPTUAL DEVELOPMENT	GRADE LEVEL	EMR	TMR	ED	LD
Attribute Games and Problems	1–3	x			
	4–6	x		x	x
	7–9	x		x	x
	10–12	x			
Geo Blocks	1–3	x	x	x	x
	4–6	x	x	x	x
	7–9	x	x	x	x
	10–12	x	x	x	x
Mirror Cards	1–3	x			
	4–6	x			
	7–9		x		
	10–12		x		
Pattern Blocks	1–3	x	x		x
	4–6	x	x		x
	7–9		x		
	10–12				
Tangrams	1–3	x	x	x	x
	4–6	x	x	x	x
	7–9	x	x	x	x
	10–12	x	x	x	x
Tracks	1–3	x			
	4–6	x		x	x
	7–9	x			
	10–12		x		

Continued on page 268.

Table 9-5 (Continued)

UNITS FOR PSYCHOMOTOR DEVELOPMENT	GRADE LEVEL	EMR	TMR	ED	LD
Clay Boats	1–3	x		x	x
	4–6	x		x	x
	7–9	x		x	x
	10–12			x	x
Drops, Streams, and Containers	1–3	x			
	4–6	x		x	x
	7–9	x		x	
	10–12	x			
Ice Cubes	1–3			x	x
	4–6	x		x	x
	7–9	x		x	x
	10–12	x		x	x
Mapping	1–3			x	x
	4–6	x		x	x
	7–9	x		x	x
	10–12	x			x
Mystery Powders	1–3			x	x
	4–6	x		x	x
	7–9	x		x	x
	10–12			x	x
Primary Balancing	1–3	x		x	x
	4–6	x		x	x
	7–9	x		x	x
	10–12			x	x
Sink or Float	1–3		x	x	x
	4–6	x	x	x	x
	7–9		x		
	10–12		x		

findings are presented on gains in the skills of observing, inferring, and communicating experienced by the experimental groups. The guide provides an abundance of directions and alternatives for special education teachers and exciting new avenues for special children.

ESS Phases of Instruction

S—APA has its instructional hierarchy or sequence of prerequisites. SCIS has its three-stage strategy of exploration, invention, and discovery. But ESS? Its phases of instruction are more difficult to describe.

The developers of ESS believed that more learning occurred through first-hand exploration chosen and directed by the learners—the types of behaviors that border on play—than through activities chosen and directed by the teacher. They believed that children should have an abundance of opportunities to absorb experiences, to exercise their curiosity according to their own abilities and interests, and to share discoveries. Can such unstructured emphases be formed into an instructional strategy?

The best attempt was described by David Hawkins, a former director of ESS, in "Messing About in Science" (1965). He discusses three phases: the *messing-about* phase, the *multiply programed* phase, and the *discussion-sharing* phase.

How might these three phases be implemented? As an example, consider the ESS unit "Ice Cubes." A teacher might provide the children with a variety of shapes and sizes of ice cubes and pose a broad question, "What can you tell me about the melting rate of an ice cube?" The children are then allowed to "mess about" to pursue answers to the question. The duration of this phase is dependent upon the students' motivation and ability as well as the teacher's patience and guidance.

The next phase is also dependent on the teacher's ability to be sensitive to the children's interests and abilities. During the multiply programed phase, a wide variety of choices should be available. The choices may be suggested either by the teacher or by the children. For example, a question generated by the teacher which lends itself to multiply programed activities might be, "What's the best way to keep an ice cube from melting?" The children might then have access to any of the following materials, along with a new supply of ice cubes:

Various kinds of papers
Rags
Soil
Sand
Vermiculite
Various kinds of liquids
Aluminum foil
Jars
Cans
Cardboard boxes

Sponges
Toothpicks
Styrofoam cups
Plastic bags
Plastic containers
Rocks
Rubber bands
Pieces of wood
Tape
Wire

The question or questions to be investigated, and the selection of materials to be used, might be generated by the students. (This is an ideal time to use learning centers or problem cards.) Regardless of the origin of the tasks or questions to be investigated, numerous options should be available.

The multiply programed phase can gradually evolve into the discussion-sharing phase. The lead question might be, "What are your results?" Discoveries can be shared; similar approaches might be compared; data can be analyzed and questioned as results are examined. This discussion-sharing phase might then lead into either additional messing-about phases or new multiply programed activities.

There is no lockstep sequence for these three phases of instruction: they may be used in *any* order. The one which is most often slighted is the messing-about phase; yet this is the experience which best typifies the ESS spirit and its attempt to respond to the way children actually learn. It may conflict, however, with various teachers' styles of instruction.

Selecting Elementary Science Curricula

Every curriculum program—not just S—APA, SCIS, ESS, or other science programs—has its strengths and weaknesses. Also, "beauty is in the eye of the beholder"—what one teacher may consider an asset another might consider a liability. The highly structural sequence of S—APA may provide invaluable guidance or ineffective rigidity. The content emphasis of SCIS may be appropriate or inappropriate in certain settings. The opportunities for open-ended, personal discoveries from ESS units may promote creativity or chaos. The strategies and expectations of the program developers may be realistic or unrealistic for different situations. Reliance on prepackaged kits can save the teacher time in organizing materials but could require more time for management problems. The information in the teacher's guide can be interpreted as suggestions to be tried and modified or mandates to be followed blindly.

Science programs are not likely to be *inherently* effective or ineffective, good or bad, useful or useless. Their success or failure depends to a great extent upon interactions between teacher and students. But because the teacher is the primary selector or designer of learning encounters, the ultimate success or failure of a science program (or even of a specific encounter) hinges on the teacher's decisions. Certainly, other variables are involved (some will be considered in Chapter 10), but *your* influence will have the greatest overall significance in science with children.

Suppose that the school board (or superintendent) has decided that this is the year to select a new elementary science program, and you are appointed to the selection committee. What questions will you and the other members of the committee need to consider? Or suppose you are assigned to a certain grade level, and the principal explains, "Here are five science programs; you decide what you want to use." What would be the basis for your selection? Perhaps you are given only a list of science topics for your particular teaching situation—how will you handle them? Imagine a discussion during a faculty meeting about the need for a different science program. What would you suggest? Are these unreal situations? Not really. But in case you're not convinced, consider an extremely likely possibility: You're told, "Here's your science program; complete as much of it as you can. Any questions?" How will you respond? You'll need to be ready with the three interrogatives previously considered—*what, why,* and *how.*

Forming a Philosophy

Selecting, building, improving, or implementing an elementary science curriculum must stem from a fundamental philosophy, or system of beliefs. You must have a basis for any of your instructional decisions—from selecting an activity for a 20-minute lesson to helping to select a science program for an entire school district. Too often this initial step is slighted or even omitted, and decisions might be made on the basis of whims or narrow preferences. Whether the decision is to be made individually or cooperatively, the following questions need to be considered.

What is your definition of science? What sort of science do you value for children? Should the acquisition of content be the major consideration? If so, should a clear conceptual framework be evident throughout the program? (Some examples of conceptual schemes or principles are: "All matter is composed of units which vary in size and complexity"; "Organisms adapt to their environment in order to survive"; and "The earth is constantly changing." A program might use a spiral approach to incorporate such ideas systematically: each year children would experience additional evidence of each principle.) Or is an emphasis on content more important in itself than a conceptual framework or a sequence of concepts? How will you judge the appropriateness of content?

Should the processes of science be emphasized more than content? Why or why not? Which processes or thinking skills are most appropriate at each grade level? How might an emphasis on processes influence the selection of things to be manipulated and types of encounters? How might classroom organization and management help or hinder the development of children's thinking skills? To what extent should process and content receive equal attention?

How important are the affective aspects of science? What types of attitudes should a science program foster? How—explicitly or implicitly—should these values be identified? Or should science encounters be value-free? Would they then become valueless? How might the development of scientific literacy be incorporated into the process of valuing? To what extent should the seven values of science which were cited in Chapter 1 be incorporated into a science program?

What is your view of learning? How does it influence your view of teaching? To what extent should a science curriculum reflect these views?

How important is science in terms of the total school curriculum? Should it be placed in the category of "extracurricular activities," or is it as basic as the three R's? To what extent are teachers obliged to meet the wishes of parents who express an interest in science for their children? (For example, 83 percent of 910 parents I surveyed in a Midwestern suburban school district considered science to be as important for their children as reading, writing, and arithmetic; but only 41 percent of fifty-six teachers and 33 percent of six principals agreed with them.) What is a teacher's responsibility for fostering the sense of wonder in children or for cultivating their natural inclination toward "scientific stuff" as described in Chapter 1? On the basis of your values, what do you want a science curriculum to accomplish?

These questions are difficult enough for just one person to answer; but the difficulty is compounded when some realistic, practical consensus must be reached by a committee. We all perceive differently. Diversity exists even within a faculty that has taught together for many years. These questions are also difficult because they cannot be answered simply in terms of the teacher's interests—they must also be answered in terms of *what will be best for the students*. But only after they have been answered—only after the basic foundation has been established—should the next step in the process of selecting science curricula be considered.

Establishing Priorities

Perhaps this section should be subtitled, "Facing Reality." Which questions from the previous section do you consider most important and most worth answering? Which questions still need to be considered? What characteristics must be present in a science program for your children?

This is a point when the brainstorming technique can be very useful. Simply list, nonjudgmentally, as many as possible of the characteristics you would like to see in a science program for children. As a basis for comparison, you can use the following list, which was volunteered by a group of elementary teachers:

Interests students
Has durable materials
Stresses creativity
Requires little training of the teacher
Is interdisciplinary
Has good materials
Balances content
Has a clear teacher's guide
Is open-ended

Is flexible
Is easily maintained
Asks good questions
Requires answers to be derived from doing, not just reading
Is easy to implement
Requires minimal reading by students
Emphasizes recording data
Takes up current topics
Avoids controversial topics
Has appropriate content for children's developmental levels
Is student-centered, not teacher-centered
Is easily individualized
Emphasizes process
Uses only metric measuring
Has easily inventoried materials
Makes good suggestions for extra projects
Develops good attitudes
Has a variety of approaches
Suggests methods of evaluation
Has reading materials for students
Has a good glossary and index
Is inexpensive
Appeals to the community
Provides a basic understanding of content
Develops good study habits
Uses clear procedures
Has appealing pictures
Has a durable binding
Has a clear table of contents
Stresses inquiry
Stresses thinking
Has a good sequence of concepts

What other characteristics do you think should be added to the list? Where do inconsistencies exist? How might these be resolved?

Try to determine which characteristics are most important; of these, are some more important than others—that is, does some sort of priority exist? Is each of the characteristics clearly understood? Would operational definitions improve communication? After comparing all the characteristics, can you decide that some are *least* important? If so, the remaining items can be used in the next step.

Developing an Analysis-Evaluation Instrument

A checklist or coding format similar to those described in Chapter 8 can be developed for analyzing and evaluating elementary science programs. Select the most appropriate items from your priorities as the criteria. Try to avoid ambiguity and items which would be difficult to identify. Various ratings could be developed: for example, a point system such as 0–5; "grades" such as A, B, and C, or percentages; choices such as "yes or no," "present or absent," "high or low"; or a continuum, as shown below:

Poor_____Excellent

A sample instrument is presented in Table 9-6. This was developed by an elementary science committee in Kirkwood, Missouri (of which I was a member), after several brainstorming sessions to establish priorities. The instrument was used during a six-month period while seven elementary science programs were examined and evaluated. The fifth category of this sample format has to do with the notion of trial testing—that is, using the program in classroom settings. Sometimes this can be done only in a few classrooms or only at certain levels; but invaluable information can often be derived from even very limited trials. Although a program might look beautiful on paper, problems may become evident during its actual use in classrooms. Without data from field testing, many of the evaluative ratings will be based on subjective inferences rather than objective observations. Additional information can be gained by observing how the program is being used in schools where it has already been implemented. Veteran teachers can provide helpful suggestions and reactions.

After a philosophy and priorities have been established, and an analysis-evaluation instrument has been developed and applied, a tentative selection can be made. But the process of analyzing and interpreting elementary science curricula certainly does not end here. Rather, this marks the beginning of two more major considerations: (1) The planning skills needed for testing or implementing the science curriculum must still be considered; since these merit more thorough analysis, they will be presented in Chapter 10. (2) Some of the factors which influence science curriculum development in general should also be analyzed; these are discussed in the following section.

Identifying the Ongoing Variables

One of the most exciting aspects of the process of developing and selecting elementary school science curricula is its similarity to the scientific enterprise in general. Both are continual, never-ending endeavors. Both represent attempts to search for meaningful explanations. The products of both are constantly being reviewed, refined, or replaced as new discoveries and better means of communication come along. And neither is ever finished. If "the best elementary school science program for all time" is ever announced, beware. There are numerous variables which make finality impossible. Some of the more obvious ones are examined below.

Demands and Influences of Society   Each decade seems to have its unique characteristics and pressures which influence the status of education and the development and selection of curricula. Recently, inflation has certainly affected the costs of science materials. Decreases in population as well as in financial support at just about every level have caused layoffs of teachers and support staff. The number of teacher aides, and especially of science supervisors and

Table 9-6

FORMAT FOR EVALUATING ELEMENTARY SCIENCE PROGRAMS

CRITERIA	PROGRAM							
1.0 In terms of the LEARNER, the science program provides for								
1.1 Active learning through the process skills								
1.2 Relevant topics, conceptually attainable (Piaget)								
1.3 Open-endedness and creativity								
1.4 Wide ranges of abilities								
1.5 Affective development (responsibility, concern)								
2.0 In terms of the PROGRAM itself, it has								
2.1 Balance among the areas of science								
2.2 Flexibility								
2.3 Sequence building on previous learnings								
2.4 Appropriate reading materials for students								
2.5 Metric measuring activities								
3.0 In terms of the THINGS TO BE MANIPULATED, they are								
3.1 Attractive to the learner								
3.2 Durable								
3.3 Easily stored and inventoried								
3.4 Safe								
3.5 Necessary for achieving the program goals								
4.0 In terms of the TEACHER, the program has								
4.1 Meaningful objectives								
4.2 Suggestions for evaluation								
4.3 Clear procedures								
4.4 Background Information when needed								
4.5 Emphasis on guided discovery								
5.0 After direct use of or exposure to the science program								
5.1 The reactions of the children were								
5.2 The reactions of other teachers were								
5.3 The reactions of the administrators were								
5.4 The reactions of the parents were								
5.5 Our overall reactions were								
TOTAL								
AVERAGE								

Scale: Blank = not applicable; 1 = low; 2 = medium; 3 = high; 4 = very high.

consultants, has decreased noticeably. Elementary teachers are expected to "cover" more topics with more children while being given less assistance. Teachers are far less mobile than they have been in the past, and this means that the likelihood of stagnation is greater. Yet the purpose and importance of in-service training—one way for teachers to keep abreast of current trends—are uncertain. Some teachers consider in-service training an imposition on their limited time; others want such training but are denied it because their administrators lack interest or funds.

Society has been demanding accountability and asking, "Are we getting our money's worth?" Many people are urging schools to move back to basics. Such demands will surely influence the development and selection of curricula; but it's difficult to predict how, especially since society has not operationally defined the terms *accountability* and *basics*. How will these pressures be felt in the next decades? What new demands can we anticipate from society, and how will these influence your decision regarding the selection of science curricula?

Research in Psychology and Pedagogy It has taken years for Piaget's ideas to filter into some classrooms and curricular materials. We've begun to learn more about the appropriateness and inappropriateness of certain concepts for children, about effective and ineffective ways to present learning encounters. But we are still faced with many unknowns. What learning styles do different children display? What teaching styles do certain teachers use? What attempts will be made to match these variables? How do our attempts to develop more logical patterns of thought help or hinder the development of creativity? What role will technology (especially computers) play in classrooms of the future? What role should it have? In the coming years, should science curricula become more interdisciplinary or more distinct and specialized? How might developments in educational research influence your decisions in the classroom?

Advances in Science Education Some children are still using science texts that end with "Some day man will set foot on the moon." What is the school's, the community's, or the teacher's responsibility for including current scientific developments in the science program?

What is our responsibility for presenting a systematic and balanced view of science to children, especially young children? In what ways might young children benefit from their encounters with science? There is an abundance of evidence from research indicating improvements in children's (1) reading readiness, (2) thinking skills, (3) mathematical skills, (4) language development, (5) problem-solving skills, (6) creativity, (7) curiosity, and (8) IQs. Some of these improvements are even more pronounced among minority or disadvantaged children. (Ten references listed in the Suggested Readings substantiate these points.)

Since there is no way of knowing with certainty what specific science content a child will need in 20 or 30 years, what general knowledge and skills should be stressed now? What additional advances in science and in science education might be reflected in the development and selection of curricula in the future?

Today's Children My personal experiences in elementary school classrooms, along with conversations with hundreds of elementary teachers across the country, support the idea that it is more difficult to educate children today than it was 15 or 20 years ago. This is not nostalgia for the "good old days" (which certainly had their not-so-good aspects) or a belief that the human race is deteriorating. Some of today's children are brighter than yesterday's. Many are far more sensitive to environmental concerns. Many have had much greater exposure to other places, cultures, customs, and accomplishments. Many have had—and benefited from—numerous opportunities which were unavailable to earlier generations. But children of today also have more serious pressures and problems than we had as elementary school children. There are great differences between the children 20 years ago and today's in terms of the use of time—specifically, the amount of time spent watching television and how time spent *not* watching television (which is very limited) is used. Children today have more chances to experiment with drugs and alcohol, participate in acts of vandalism, be alone (even lonely) or unsupervised, or be victims of sexual or physical abuses. It should come as no great surprise, therefore, that many are restless, passive, uninterested, bitter, or easily distracted and impatient for quick, simple solutions. Many have been conditioned to expect constant comfort and immediate gratification—partly as a result of their parents' attitude: "We want our kids to have it better than we did."

Disequilibrium is still necessary if learning is to occur. But this component of the learning process is increasingly unknown, denied, or avoided by today's children, and perhaps by many of their parents. How will you attempt to provide encounters which elicit disequilibrium, intervene carefully, and yet allow children to adapt (that is, learn) by resolving disequilibrium themselves? More general questions are: How might the elementary school science curriculum respond to the needs and interests of today's children? How might tomorrow's children compare with today's? What additional adjustments in science curricula may have to be made to meet the needs of tomorrow's children?

Additional Considerations on Selecting Curricula

The analyses of S—APA, SCIS, and ESS and the procedures for selecting elementary science curricula given in this chapter are only a sampling of the many variables you must be prepared to consider. When you analyze new elementary science textbooks or science programs other than textbooks, you will discover the significant contributions of the "alphabet curricula." (Some publishers of elementary science textbooks are listed in the Suggested Readings.) Look especially for the developers' attempts to balance process, content, and personal experience and to balance biological, physical, and earth sciences. Don't look for easy solutions or prepackaged answers. Instead, try to identify, analyze, and interpret the many variables present—the type of learning you hope to make possible (the responding variable), the factors that you can try to control, and the choices you hope will facilitate meaningful experiences in science for your students. As for the many uncontrolled variables, the planning skills discussed in Chapter 10 can lessen their influence.

Table 9-7

	S—APA	SCIS	ESS
Program components			
Learning emphases			
Teaching strategies			
Strengths			
Weaknesses			

Encounters

1. Determine which schools in your area are using S—APA, SCIS, or ESS. Visit any of the schools to observe the use of these programs. What are the teachers' attitudes toward the programs? What are the children's reactions?
2. If possible, examine some actual modules and units from S—APA, SCIS, and ESS. Construct a chart for comparing the three programs. Use the format shown in Table 9-7 or any modification which suits your interests.
3. Examine a variety of elementary science textbooks. Determine the most significant differences and similarities between textbooks programs and the "alphabet curricula."
4. What do you think are the most important factors to be considered in selecting an elementary school science program? Interview your colleagues, other classroom teachers, and some elementary school children to discover their views. Then analyze your findings. How have the additional insights influenced your original choices?
5. Design a checklist (or any suitable format) for evaluating elementary school science programs. Use it to evaluate several programs and share your results with others. Which factors elicit the most agreement? The most disagreement?
6. Of the ongoing variables described in this chapter, which do you think have the greatest influence on elementary school science curricula? How are today's elementary science programs meeting the needs and interests of today's children? What are your predictions for the development and selection of science programs for tomorrow's children?
7. Your choice.

References

Aho, William, Del Alberti, Victor Perkes, Robert Sheldon, Terry Thomas, and Richard Ward: *The McGraw-Hill Evaluation Program For ESS*, McGraw-Hill, New York, 1974.

Ball, Daniel W.: *ESS/Special Education Teacher's Guide*, McGraw-Hill, New York, 1978.

Guide for Inservice Instruction: Science—A Process Approach, American Association for the Advance of Science, Washington, D.C., 1973.

Hawkins, David: "Messing About in Science," *Science and Children*, National Science Teachers Association, Washington, D.C., vol. 2, no. 5, February 1965.

Suggested Readings

For Additional Information

To request additional information about the "alphabet curricula" described in this chapter, write to the following addresses:

Elementary Science Study
>Webster Division, McGraw-Hill Book Company
>1221 Avenue of the Americas
>New York, New York 10020

Science–A Process Approach
>Ginn and Company
>191 Spring Street
>Lexington, Massachusetts 02173

Science Curriculum Improvement Study (SCIIS)
>School Department
>Rand McNally & Company
>P.O. Box 7600
>Chicago, Illinois 60680

Science Curriculum Improvement Study (SCIS II)
>American Science and Engineering, Inc.
>200 Overland Street
>Boston, Massachusetts 02215

Additional "Alphabet Curricula"

Conceptually Oriented Program in Elementary Science (COPES)
>American Science and Engineering, Inc.
>200 Overland Street
>Boston, Massachusetts 02215
>(COPES is a nontextbook, activity-oriented science program for grades K–6 with teachers' guides centering on five conceptual schemes: structural units of the universe, interaction and change, conservation of energy, degradation of energy, and a statistical view of nature. The program does not rely heavily on kits, since most of its materials can be obtained locally.)

Minnesota School Mathematics and Science Teaching Project (MINNEMAST)
>Minnemath Center
>720 Washington Avenue, S.E.
>Minneapolis, Minnesota 55414

(MINNEMAST is a program for grades K–3 consisting of twenty-nine sequential units which coordinate science and mathematics. Fifteen of the teachers' guides deal with science topics and the development of the thinking skills. "Watching and Wondering," "Describing and Classifying," "Using Our Senses," and "Introducing Symmetry" are excellent enrichments for any kindergarten program.)

Unified Science and Mathematics for Elementary Schools (USMES)

Education Development Center, Inc.

55 Chapel Street

Newton, Massachusetts 02160

(USMES is a collection of teachers' resource guides applicable for grades K–8. The project integrates science, mathematics, social studies, language arts, and career education and provides opportunities for children to develop problem-solving skills. The unit topics deal with real-life situations—for example, "Pedestrian Crossings," "Consumer Research," "Designing for Human Proportions," and "Bicycle Transportation"—and present practical challenges to be investigated by the students.)

Articles

The following articles indicate how children in grades K–6 have benefited from science encounters which involve their interacting with materials:

Barufaldi, J.P.: "Children Learning to Read Should Experience Science," *Reading Teacher*, vol. 30, pp. 388–393, January 1977.

Brown, W.R.: "A Look at Integration of Science and Mathematics in the Elementary School," *School Science and Mathematics*, vol. 76, pp. 551–562, November 1976.

Davis, A.R.: "Science for Fives," *Childhood Education*, vol. 53, pp. 206–208, February 1977.

David, T.: "Comparison of Achievement and Creativity of Elementary School Students Using Projects versus Textbook Programs, *Journal of Research in Science Teaching*, vol. 13, pp. 205–212, May 1976.

Harbeck, M.B.‘ "Is Science Basic? You Bet It Is," *Teacher*, vol. 94, p. 22, November 1976.

Herron, J.D.: "Implicit Curriculum: Where Values Are Really Taught," *Science Teacher*, vol. 44, pp. 30–31, March 1977.

Morgan, A.: "Sciencing Activities as Contributors to the Development of Reading Skills in First Grade Students," *Science Education*, vol. 61, pp. 135–144, April 1972.

Rowe, M.B.: "Help Is Denied to Those in Need," *Science and Children*, vol. 10, pp. 23–25, March 1975.

Smith, R.F.: "Science: A Lever for Primary Grades," *Childhood Education,* vol. 53, pp. 203–204, February 1977.

Watson, F.G.: "Science: The Basic," *Science Teacher*, vol. 44, p. 19, March 1977.

Some Publishers of Elementary Science Textbook Series

Addison-Wesley Publishing Company
School Division
Sand Hill Road
Menlo Park, California 94025

Allyn and Bacon, Inc.
Rockleigh, New Jersey 07647

American Book Company
300 Pike Street
Cincinnati, Ohio 45202

Benefic Press
10300 West Roosevelt Road
Westchester, Illinois 60153

Benzinger, Bruce, Glencoe
The Macmillan Company
Riverside, New Jersey 08075

Bobbs-Merrill Company
4300 West 62 Street
Indianapolis, Indiana 46268

Follett Publishing Company
1010 West Washington Boulevard
Chicago, Illinois 60607

Ginn and Company
Statler Building
Back Bay, P.O. 191
Boston, Massachusetts 02117

Harcourt, Brace, Jovanovich
757 Third Avenue
New York, New York 10017

Harper and Row, Publishers, Inc.
Keystone Industrial Park
Scranton, Pennsylvania 18512

D. C. Heath and Company
2700 North Richardt Avenue
Indianapolis, Indiana 46219

Holt, Rinehart and Winston, Inc.
383 Madison Avenue
New York, New York 10017

Houghton Mifflin Company
110 Tremont Street
Boston, Massachusetts 02107

Laidlaw Brothers
Thatcher and Madison Streets
River Forest, Illinois 60305

J. B. Lippincott Company
East Washington Square
Philadelphia, Pennsylvania 19105

The Macmillan Company
866 Third Avenue
New York, New York 10022

McGraw-Hill Book Company
Webster Division
Manchester Road
Manchester, Missouri 63011

Charles E. Merrill Books, Inc.
1300 Alum Creek Drive
Columbus, Ohio 43216

Rand McNally and Company
School Department, Box 7600
Chicago, Illinois 60680

Random House—Singer School Division
201 East 50 Street
New York, New York 10022

Scott, Foresman and Company
1900 East Lake Avenue
Glenview, Illinios 60025

Silver Burdett Company
250 James Street
Morristown, New Jersey 07960

CHAPTER 10
TESTING YOUR DECISION-MAKING SKILLS

ARE YOU READY? PAUSE BEFORE PLUNGING

All the previous chapters have been leading up to this one. All your observing, analyzing, and interpreting of children's behaviors were prerequisite experiences. The same is true of your experiencing the processes of science, observing classroom interactions, and analyzing elementary science curricula. A synthesis of all these preliminaries is essential for planning science with children in the real world of the classroom.

Planning is a very familiar notion. We all plan, for example, when important personal decisions are concerned. Who would decide to move to a new city, embark on an extended trip, invest one's savings, select a new career, or enter into any new living arrangement without some degree of planning? The unprepared teacher may survive in the classroom for a short period—but not for long. There are too many factors involved in teaching and learning that cannot be left to chance. Even though we do not cause learning, we are professionally responsible for deciding which conditions should be the most conducive to learning.

If you hope to find or be given a foolproof set of decision-making skills, you'll be disappointed. There will be no magic formula presented in this chapter which will transform you into an effective decision maker. If you accept this fact, I hope you will also accept its implications. Since no single format or strategy can meet the needs, styles, or purposes of *all* teachers for *all* learners, you have the freedom to pick and choose your own. You have the opportunity to organize your best predictions and hypotheses on teaching and learning. No teaching machine, and no other person, will ever be able to organize your own hunches and decisions as well as you can. The chance that any developers of science curricula have written materials specifically for you and your students is remote. Only you can do this. It is, of course, worthwhile to assimilate the useful ideas of others; but it is more fulfilling to adapt them to make your own personal expressions. Because there are no definitive procedures (and any presentation of "*the* only way" should be held suspect), you have an unending task of decision making. You have the daily responsibility of sifting, sorting, and deciding—deliberately and with discrimination—to come up with what you

283

hope will be meaningful science encounters for children. You've had this same responsibility throughout this book. But the types of analysis and synthesis incorporated in this chapter necessitate even greater scrutiny and evaluation. Like all the processes presented so far, the development and use of decision-making skills are continuous. The more you test them, analyze their results, and evaluate their effectiveness, the better you will be able to determine what works best for you and your students.

In planning science encounters for children, the fundamental considerations are quite similar to those involved in planning a trip. Both situations require answers to the same basic questions:

Where are we?

Where are we heading?

How do we get there?

How will we know when we've arrived?

Although questions are similar for traveling and teaching, the search for answers is quite different in each—and, as you will discover, more difficult in teaching.

Where Are We, and Where Are We Heading?

These first two questions have been combined to avoid the egg-and-chicken debate. Which needs to be answered first? In all likelihood, they are of equal importance. The question "Where are we?" is often ignored, however, or considered insignificant by some teachers—even though (to return to the analogy with traveling) if you don't know your present location, it's impossible, or at best extremely difficult, to reach your intended destination.

When the implications of Piaget's ideas were presented, a similar question was asked—"Do you know where your children are?" This question has to do with identifying students' *entry behaviors* before any specific intended instruction. Your experiences with the Piagetian-type interviews probably have convinced you that a class of, say, twenty third-graders may well represent twenty different kinds of intellectual development. These children probably also differ significantly in emotional (or affective) development as well as in manipulative (or motor) development.

When you ponder the multitude of facts, concepts, principles, skills, and opportunities for expressions of attitudes that *could* be presented to children, it's almost overwhelming. When you attempt to answer the question "Where are we heading?" try to decide on a *specific* end product or observable behavior. A statement of the teacher's intention which describes a desired change in the learner's behavior is generally called an *instructional objective*, *performance objective*, or *behavioral objective*. Your decisions about each of the major components of an instructional objective can provide valuable answers to the question "Where are we heading?"

Specifying Instructional Objectives

Instructional objectives express what the teacher hopes the learners can accomplish as a result of instructions or encounters presented within a particular learning environment. As operational statements of expectations, they are *means* (not ends). They enable the teacher to attempt to answer four very important questions:

1. What should the learners be able to do as a result of their interactions in the learning encounter?
2. What materials and procedures will be most conducive to the desired behaviors?
3. What degree of mastery of learning should the learners demonstrate?
4. Why should the learners be expected to demonstrate the desired behavior?

Restated in terms of the teacher, the four questions become:

1. What should I teach?
2. What conditions are needed to teach what I want to teach?
3. How will I know when I've taught it?
4. Is it worth teaching?

These two sets of questions illustrate two different emphases and uses of instructional objectives. The first set emphasizes the view that learning occurs within the learner; the teacher designs and facilitates learning encounters. In the second set, greater emphasis is placed on the teacher's actions; learning is considered primarily as a response to stimuli. Regardless of the teacher's view of the teaching-learning process, then, instructional objectives can still be formulated.

Both sets of questions relate to four major components of instructional objectives: (1) the learning condition or situation; (2) performance or behavioral terms; (3) standards for performance; and (4) the rationale for the objective. When instructional objectives are constructed as formal statements, the learning conditions are usually stated first. (This is not a rule, but it is common practice.) Otherwise, the performance terms are stated first.

The Learning Condition or Situation   The first component of an objective is the situation in which the learners are expected to demonstrate certain performances. It includes the physical environment at the time the learning is to occur—that is, the instructional conditions and the materials to be used. The description of the learning condition may contain accounts of previous instructions. Generally, however, only the immediate situation needs to be described. The following are examples of possible learning conditions or situations for instructional objectives:

Given a diagram of the human skeleton, . . .

Given a birthday candle, a match, a small piece of clay for supporting the candle in a vertical position, a metric ruler, and paper and pencil, . . .

After constructing a mini nature trail and observing those constructed by others, . . .

There's no set formula for specifying learning conditions. Often such a specification will look like a "grocery list" of materials. The main criteria are discretion and clarity. Are these the best materials and conditions for determining whether the desired behaviors have been learned? Is the description clear—that is, could another person (such as a substitute teacher) reproduce the same conditions in your absence?

Performance or Behavioral Terms   This component specifies the actions the learners are expected to perform as a result of interacting with the learning condition. One of the major purposes of formulating instructional objectives is to enhance communication: it is imperative that the performance term should designate actions from the learners which are observable and measurable, that is, readily recognizable. If it does not, communication will be hindered. The following are examples of actions which can easily be observed:

Analyze
Arrange
Classify
Compare
Construct
Demonstrate
Describe
Design
Evaluate
Identify
Infer
Interpret
Hypothesize
List
Measure
Organize
Select
Summarize

By contrast, it's extremely difficult to recognize the following behaviors; in particular, it is hard to recognize when a learner is *failing* to demonstrate them:

Know
Understand (or "develop an understanding of")
Appreciate (or "fully appreciate")
Grasp the significance of

Have a strong faith in
See the necessity for

These are certainly worthwhile goals, but they are virtually impossible to recognize unless they are operationally defined. For example, a teacher would need to specify what should be observed when a child is understanding or appreciating.

One of the easiest ways to determine if a performance term is observable and measurable is to apply Robert Mager's "Hey, Dad" test from his book *Goal Analysis* (1972, pp. 29–31). Simply add the selected performance to the end of the statement, "Hey, Dad, let me show you how I can . . ." Then envision the results. "Hey, Dad, let me show you how I can *describe* these objects" can be clearly visualized; but "Hey, Dad, let me show you how I can *develop an appreciation* of these objects" cannot. A student either can or cannot describe the objects. But it's extremely difficult to recognize when one has developed an appreciation of them.

The performance term is extremely important, and certainly the pivotal component of the instructional objective. It is the major focus or direction of the teacher's intentions. The performance term provides the answer to the questions "What should the learners be able to do as a result of interacting with the learning conditions?" and "What should I teach?" The merit of the learning condition can be determined only in relation to the performance term. This is also true of the next component, the standard.

Standards for Acceptable Performance   This component states how well or to what degree the learners are expected to perform a desired task. In other words, it provides answers to the questions "What degree of mastery should the students demonstrate?" and "How will I know when I've taught what I intended to teach?" This component can often be the most difficult to determine, especially for a neophyte teacher or even for a veteran teacher facing a new group of learners. What are reasonable, acceptable standards of performances? Where is the line drawn between expectations which are too lenient and expectations which are too demanding, or between satisfactory and unsatisfactory performances? Should all learners be expected to demonstrate an identical degree of mastery?

In one of his early programed books, *Preparing Instructional Objectives*, Robert Mager tended to concentrate on quantifying minimal acceptable performances (1962, pp. 44–52). That is, the acceptance level could be stated in terms of quantities, such as the following:

Number of correct answers or actions (fifteen out of twenty)
Percentage of correct answers or actions (75)
Time limit within which a skill should be demonstrated (10 minutes)
Number of times an action is attempted (at least five)

In certain areas of learning, an emphasis on quantities may be valuable, particularly if speed and accuracy are necessary (examples might be hospital work, typing, or athletics). However, these attributes are generally not as important for the teaching and learning of science at the elementary level. In fact, sometimes specifying standards in terms of quantities can border on absurdity. Consider the following example:

Given a metric tape, pencil, and data sheet, the students should measure the dimensions of objects in the classroom. The minimal acceptable performance consists of 78 percent of the students measuring fifteen objects, with measurements accurate to within 5 mm, during a 10-minute period.

Although it may sometimes be more difficult to specify, the *quality* of the students' performance is usually far more important than the quantity. When you attempt to analyze quality, you will actually be operationally defining what you hope will occur as a result of the teaching and learning. Consider the following example, which incorporates standards of both quantity and quality:

Given a birthday candle, a match, a small piece of clay for holding the candle in a vertical position, a metric ruler, and pencil and paper (the *learning conditions*), the learners should be able to observe and record observations of the candle before, during, and after burning (the *performance terms*). Acceptable performance consists of a minimum of fifteen observations within 10 minutes (*quantity required*). The observations should include data from all five senses, measurements of physical properties, and descriptions of changes that have occurred (*quality of performance required*).

Another way of expressing a standard for performance is as a position or positions on a continuum:

Minimum performance_____Maximum performance

A continuum is useful if you wish to specify and identify expectations for students with differing abilities. For example, what might you consider as acceptable performances from the least capable students? From the average students? From the most capable students? There are a few skills that all students should be able to demonstrate rather uniformly, such as measuring with a meter stick, reading a thermometer or graduated cylinder, and using a balance. But most learning is acquired and subsequently demonstrated in terms of *degrees* of mastery. Seldom is a single standard for acceptable performance appropriate for *all* learners.

Rationale   This component is the heart of instructional objectives. It is the answer to the questions "Why should I teach this?" and "Why should the learners be expected to learn this?" In other words, it's the justification of the expectations.

The rationale may or may not be stated explicitly in the written objectives of the teacher or those printed in curriculum guides. But it must be known by the teacher and communicated to the learners. In fact, one of the toughest jobs a teacher constantly faces (besides selecting the learning encounters) is to have *realistic* answers ready for questions like "Why do I have to learn this?" and

"Why do I have to do this?" The days are long gone when the teacher could answer with the authoritative "Because I said so" or the condescending "Because it will help you when you grow up." For too long, curriculum developers and teachers have been dishing out ideas or tasks which, for the most part, students have had to take on faith. Educators have often expected students to fit the course objectives, rather than developing objectives to fit the students. Common sense, our own past experiences, and sound pedagogy, however, should tell us that any learner is more apt to pursue goals that are of immediate use or have personal value. *The learner has both the need and the right to know not only what is expected but also why it is expected.*

Perhaps all this will make more sense if you think about your own experiences as a learner. How might your own learning experiences have been different had each of your teachers communicated a rationale for their expectations—clearly, honestly, and in a way that made sense to you? How might your motivation or that of your peers have been affected? If every teacher were somehow forced to consider rationales for all objectives and then communicate them to the students, how might curricula be affected? What expectations would most probably be eliminated? What would be the most likely additions?

Often, when one writes about—or simply thinks about—an instructional objective, the rationale is the last component to be considered. However, in a real-life situation it should be considered first. After considering the rationale, you will be in a much better position to judge the merit of instructional objectives in general.

Recognizing the Effects of Instructional Objectives

Like most tools, techniques, or methods of operating, instructional objectives tend to be neutral in themselves. It is our interpretation and subsequent use of them that makes them advantageous or disadvantageous, broadening or limiting. Therefore, it's impossible to formulate an accurate list of the "ten most dangerous" or "ten best" characteristics of instructional objectives. The comments in the next section are simply intended to help you become more aware of some of the weaknesses and strengths of instructional objectives.

Disadvantages   When instructional objectives are used as ends in themselves, or otherwise misconstrued, they can create certain disadvantages. Some examples follow.

Overemphasis on certain skills   The tendency has been to designate primarily those skills (state those performance terms) which are most easy to recognize and measure—for example, *name*, *describe*, *order*, *distinguish*, and so on. But an enormous number of higher cognitive skills are worth achieving which are less readily identifiable and more difficult to quantify, such as *summarize*, *analyze*, *synthesize*, *criticize*, and *evaluate*. There are also inner feelings, which are the most vital ingredients of learning—*appreciating*, *enjoying*, *yearning*, *respecting*, *wondering*, *caring*. If expectations are limited to only the most obvious skills, there is a real danger that many opportunities for widening the spectrum of thinking processes will be missed. There is also a danger that the teaching and learning of science could become mechanistic conditioning rather than an interpersonal sharing of encounters.

Stifling of interest or creativity   Sometimes if one is excessively intent on finding a particular treasure, others of equal or even greater value might be overlooked. Suppose that the objective for a lesson is "ordering a collection of minerals according to their relative hardness." But one group of children becomes more interested in the presence or absence of cleavage; another group in discovering the color of the marks left by each specimen when it is rubbed across an unglazed porcelain tile; and another group in determining whether the minerals are magnetic or not. A teacher who wears behavioral objectives like blinders would ignore these spontaneous interests and insist that the children attend solely to ordering the minerals according to hardness. But rigid adherence to a predetermined objective can squelch creativity and spontaneity. This effect is also evident when curriculum developers or test designers are allowed to dictate learning objectives. When the decisions of outside experts are accepted without question over those of the teachers and learners, the decision-making process can be impaired. The same can be said of teachers who "teach for the test" rather than guiding students toward their own discoveries.

Overreliance on objectives   Because the questions "What constitutes effective teaching?" and "How do children learn best?" may never be finally answered, the presence of instructional objectives can be an enormous temptation. Here at last is a means for identifying success or failure, for determining whether or not knowledge and skills have been acquired. Here is a panacea for those overburdened by having to decide what to do next or by the problems of evaluating. Unfortunately, we cannot rely totally on objectives no matter how much we are tempted to do so. We still know little about identifying and measuring internal responses or social interactions, or about predicting how much of what we've presented will ultimately be retained by the learners. Most educators will admit that it is impossible to specify *all* the desired outcomes of instruction. But it is misleading to view objectives as a panacea or to believe that if a performance is observable and measurable it is therefore more valuable than one that is not.

Advantages   When instructional objectives are used as *means* for communicating instructional intentions in the classroom, what potential advantages might they offer?

Increased purposefulness   If teachers would consistently attempt to answer the four questions on page 285, a great deal of purposeless effort could be eliminated. The greatest potential advantage of instructional objectives is that they impel one to make deliberate decisions. It seems reasonable to hope that the greater the conscious deliberation put into the decision-making process, the more purposeful the results, and the more reliable the answers to that most important question: "What's really worth teaching and learning in elementary science?"

One factor determining whether this potential advantage is realized is the teacher's ability to communicate instructional objectives to the learners as *commonly valued* goals. Just as children can often learn from each other, so too they can often make valuable contributions to the construction and implementation of instructional objectives. If they are allowed to do so, communication can be bilateral rather than unilateral, and the entire process of science with children can become more purposeful. An important rule might be: "If you *and the children* are not excited about the objective for the next lesson, stop and find out why before proceeding."

A basis for evaluation If the instructional objectives have been thoughtfully constructed and clearly communicated to the students, the process of evaluating performances becomes greatly simplified. By analyzing each component of the instructional objectives, you can determine the source of possible weaknesses either in your own planning or in the reactions and performances of the students. For example, to what extent was the learning situation pertinent? To what extent were the teaching materials effective? Was the expectation for performance relevant and realistic? Was the standard for acceptable performance appropriate? How well did the students understand the reason or reasons for the objective? The evaluative process used by a teacher is rather similar to that used by a physician. Both must make a diagnosis, plan and administer treatment or guidance, and then reevaluate. Carefully constructed instructional objectives can help us handle the problem of meeting the needs of the individual learner more realistically. One objective may not always serve the needs of all children; but individual objectives can be constructed, or at least the standards for acceptable performance can be modified according to the students' ranges of ability.

Facilitation of planning Rarely do teaching and learning occur in single giant steps or intuitive leaps. The hundred-yard run is rare in football and virtually nonexistent in education. No teacher can face a classroom of children and survive for long with a general goal such as "teaching all about plants," or electricity, or sound, or any natural phenomena. Goals must eventually be broken down into more manageable, meaningful components, that is, into specific instructional objectives. When the steps toward the larger goal—the specific objectives—are planned in advance, much aimless wandering can be avoided. Without such planning, a teacher can be influenced by whims or momentary pressures which may or may not be relevant to the ultimate goal.

Additional Considerations on Where We Are and Where We're Going

Once we are clear about our ultimate goal and the steps leading to it (*Where are we going?*), we find ourselves faced again with the question *Where are we?* That is, *Where are the children in relation to the goal?* Given the physical constraints of the classroom and the presence or absence of equipment, can the children achieve the goal? Also, children have all sorts of entry behaviors and existing understandings (and misunderstandings); which of these specifically relate to the goal? What individual differences exist among the children in terms of their abilities to achieve the goal? What similarities are there?

Once you develop a better understanding of where the children are in relation to the goal, you can more accurately predict the amount of time they'll need to achieve it. This prediction and the planning of specific steps toward the goal—prerequisite instructional objectives—will be influenced by the answers to the next question.

How Do We Get There?

This question is probably the most open-ended of the four. The children's entry behaviors are already present; your task is to discover and identify them. Many times the instructional goals are also given by a designated science curriculum or guidelines you're asked to follow. Sometimes objectives should be constructed according to the requests or interests of the children. Even though the answer to the question "How do we get there?" is often provided in curricular materials, you'll still need to consider the appropriateness of the suggested sequence of procedures, the proposed modes of instruction, the organization of students, and the selection and management of materials.

Sequencing Procedures: Task Analysis

The mechanics of task analysis—the process of analyzing the components of a final objective—are relatively simple. Ask the key question: "*What must the learners be able to do* in order to achieve the final objective of a topic or series of experiences?" In other words, what related learning or subordinate skills are needed? Once you have answered the key question in terms of your major goal, you will discover the first subordinate goal. The key question should then be directed to this first subordinate goal, and the answer will give you the second subordinate goal—and so on. Figure 10-1 shows the procedures for analyzing a sequence of tasks.

Keep asking the key question ("What must the learners do . . . ?") until you reach an answer that is just slightly beyond the presumed entry behaviors related to the final task. The answer to the final question will be the first instructional objective in your sequence of procedures. By applying the procedures of task analysis, then, you evolve an outline of the necessary instructional sequence. The procedure is, again, quite similar to planning for a trip. What intermediate goals (or locations) must be reached before arriving at the final goal? The greater the difficulty of the major goal in relation to the learners' entry behaviors, the more subordinate related goals there must be.

Here's an example. Suppose a fourth-grade teacher wanted the students to be able to *predict the amounts of different litter likely to be found on the*

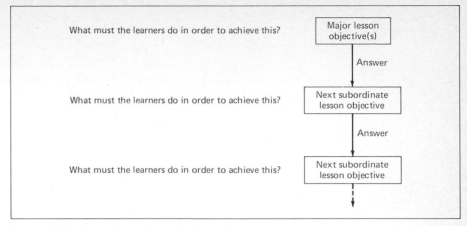

Figure 10-1 General procedures of task analysis.

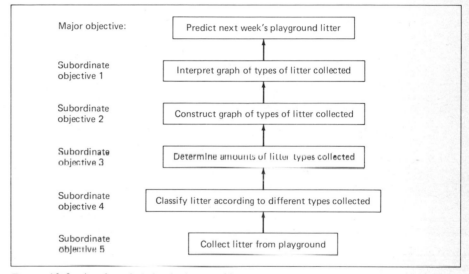

Figure 10-2 A task analysis for the litter problem.

playground next week, on the basis of a graph of the types of litter found during this week. By using the sequence of questions just described, the teacher might arrive at the task analysis shown in Figure 10-2.

Not all task analyses fit this linear sequence, with each task building on or stemming from the previous one. Sometimes a series of subordinate tasks might be at the same level of difficulty. For example, a teacher using the ESS phases of instruction might plan the task analysis shown in Figure 10-3 (page 294) for a portion of the unit "Mystery Powders."

The three-stage strategy of SCIS could also be used as a guide for sequencing a task analysis.

Once you have actually made and analyzed an outline derived from a task analysis, you're in a much better position to reconsider the learners' entry

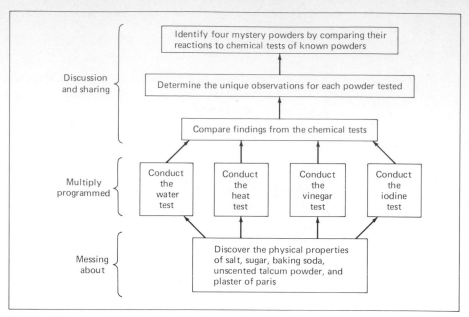

Figure 10-3 A task analysis based on the ESS phases of instruction.

behaviors and the appropriateness of your ultimate goal. But you must realize that a sequence of instructional procedures developed by task analysis represents only your best predictions. If you wish to strengthen these predictions so that your planning becomes less tentative, you'll need to go a step farther.

Pretesting is one of the surest ways of verifying your initial tentative plans or hunches. If you've attempted to answer the first three questions (*Where are we? Where are we headed? How do we get there?*), you already have the basis for a pretest. You must simply organize your thoughts and formulate your questions or tasks for the students. Let's recapitulate: Check your assumptions about the children's entry behaviors of which you are least sure. Ask a question or present a task similar to the final objective toward which you're headed. (If they already have this skill or understanding, you'll need to modify your plans.) Select any of the subordinate tasks that seem the most questionable and determine whether or not they need to be included in your instructional sequence. The time spent in pretesting can be brief, and it can save you a great deal of unnecessary instructional time and lessen frustration or boredom for the students. It will also give you additional confidence that you know not only where you're heading and how to get there but also that you're on the right track. In addition, you can proceed more confidently with the next decisions that need to be made.

Selecting Modes of Instruction

Selecting modes of instruction is one of the most exciting components of planning because it opens many avenues for personal expression and creativity. The following are just a few of the numerous possibilities from which you can choose:

Demonstration by the teacher; you're center stage, and the students watch and listen

Involvement of students with selected materials

Heavy emphasis on the use of audiovisual aids (films, overhead projections, slides, filmstrips, recordings, television, bulletin boards, pictures, trade books, resource books, magazines)

Readings by the teacher, the students, or both

Lecture by the teacher, or lecture combined with materials and demonstration

Student projects and reports

Lab sheets (dittos) for recording observations, answers, procedures

Centers for self-directed learning

Discussion

"Show and tell" by the students

Games or simulation activities

Field trips—away from, near, or on the school grounds

Self-instructional materials

Pantomimes, plays, and other dramatizations

Team teaching

Guest lecturers

Interviews, surveys, panels, debates

Your selection of any one mode or a combination of several depends upon your instructional objective for the students. Which means will best help them achieve the intended goal? You should realize, however, that the repeated use of one technique over a long period of time can eventually lose its appeal. Variety is needed to spice up science lessons, too.

Organizing Students

You'll also need to consider the advantages and disadvantages of lessons designed for the total class, for small groups, or for individuals.

Teaching must be a human act; there must be some sort of personal relationship between you and the learners. Yet, in reality, few of us can relate well to twenty or thirty people simultaneously. This makes small groups attractive. Children also have more opportunities for sharing, discussing, and cooperating (in both the social and the Piagetian sense) in small groups than as members of a large group. They're not born with the social skills necessary for cooperation; these have to be developed gradually. By participating in small-group interactions, children can gain valuable experiences needed for the development of self-control and self-discipline. Also, we cannot expect children to automatically become independent thinkers. They need to be gradually weaned from viewing the teacher as an absolute authority. By interacting with only a few of their peers, children have more opportunity for active, personal, and direct involvement in the learning process. There is also more potential for expressions of creativity and original thinking. Perhaps the greatest benefit that can be derived from small groups is the possibility that the parts can be greater than the whole.

But arranging children into small groups gives the aggressive opportunities to become even more dominating; and timid children grouped with aggressive ones become even more reticent. There are also logistical problems. If you arrange your students in five or six groups, you may often find yourself wishing you had five or six heads, ten to twelve hands, or at least bilocation. Because small-group activities generally entail greater involvement by students, they can lead to an increase in noise and distractions and create discipline problems. They also demand greater attention to collecting, distributing, and managing the learning materials.

Selecting and Managing Materials

Effective science encounters for children involve their interacting with materials—more specifically, with *real* objects from the natural world. If a choice is to be made among printed descriptions of objects, pictures, models (even lifelike models), and natural objects, select the natural objects whenever possible. There's enough artificiality in the child's world; science teachers certainly need not add more.

Selecting what you hope will be the most meaningful learning materials for your children is a crucial task. You can complete all the mental preparations for presenting a meaningful science encounter and then see your efforts fizzle because the materials you selected were meaningless or mishandled.

Either before or while you select the learning materials for your intended lesson, try to be conscious of the following characteristics:

Safety. Avoid objects that might be injurious to the children. Be prepared for breaks, spills, stains, and burns. If objects are to be tasted or smelled, first check the students' records for information on allergies or other susceptibilities. Small animals such as gerbils, hamsters, or crayfish should be handled cautiously. Avoid handling any wild or unknown animals of the "look what I found on the way to school" variety.

Support. Select objects or materials that facilitate the achievement of the lesson objective or objectives; avoid materials that distract the children or hinder the lesson.

Appeal. Pictures from affluent suburbia could be unappealing or even offensive to underprivileged children. Materials that are tattered, dirty, outdated, and generally past their prime might indicate a lack of care on your part and make the students lose interest.

Working condition. Balances should balance, microscopes should magnify; support stands should support; magnets should attract magnetic materials; wheels on axles should turn. There's enough to be concerned about in the teaching and learning of science without faulty equipment. Test such materials first.

Replacements. If possible, use material that is easily replaceable. Some items—

such as small magnets, marbles, and balloons—have an uncanny way of disappearing. Also, try to have a sufficient quantity of materials so that not many children will have to wait their turn.

Manageability and durability. Very young children generally cannot handle relatively small or large objects easily. If you spend time and energy constructing your own materials, make them strong so that they can be used more than once. Most children are not hesitant to use materials, but a few may shy away from objects that look very fragile or overly complicated.

Even after considering all these points, you have no guarantee that any particular materials will be appropriate for any particular class. Selecting materials carefully is not enough; you'll also need to consider how to go about arranging and managing them efficiently. Here are seven suggestions you might consider worth trying.

First, think about the best possible timing. When should the students actually have the materials in their hands or before their eyes? Must they have them at the very beginning of the lesson? Generally, students are more interested in interacting with materials than with spoken or printed words. So if it is important that something be said or read first, don't distribute the materials until this is done. This strategy also allows the students to give their undivided attention to the materials when they are distributed.

Second, develop any workable technique that will let you distribute the materials to the class easily and quickly. Much valuable time can be wasted if materials are distributed carelessly. You might try packaging the items, or arranging supply centers in various sections of the room, or appointing certain children as helpers.

Third, decide in advance where the children are to use the materials, and anticipate the consequences. Prepare for the inevitable—children tend to become more excited or even unruly when interacting with materials on a floor area or outdoors than when seated at desks or around tables.

Fourth, be aware that children need some initial time to investigate and handle the materials freely (within the limits of safety, of course) and independently. It's so easy to forget that these materials may be new to the children. If you rush into your planned procedures too quickly, you might stifle some potential learning.

Fifth, try to anticipate any misuses of materials, and the effects of overstimulation. If the children have had very few opportunities to interact with science materials in a classroom, their excitement over even seemingly mundane objects can be astonishing. It might be necessary to establish some behavioral guidelines before the materials are distributed.

Sixth, develop and maintain thorough cleanup procedures. You'll be doing yourself an enormous favor and also helping the children develop a worthwhile habit. You might assign a rotating cleanup detail until the children voluntarily assume the responsibility. If it's necessary to store materials, review the last "Extension" from the activity "Space and Shape" (page 115), which involved determining how many smaller boxes could fit into one larger box. Label your storage containers so that the objects can be easily identified and inventoried.

Seventh, if the science materials must be shared with other teachers, agree on the most convenient depository. Also, develop a system for determining where the materials are being used and when they will probably be returned.

Note In classrooms that are fortunate enough to have them, teachers' aides can often assist in collecting, arranging, and managing science materials. Some school districts even have a science consultant or science supervisor. This person can give you invaluable help in selecting the most appropriate materials and even in selecting procedures and modes of instruction. But very often you're on your own as you attempt to formulate, implement, and evaluate your plans.

How Will We Know We've Arrived?

Determine to what extent the students' actual performances matched the instructional objective. If there's a one-to-one correspondence, they've "arrived." If not, some may still be en route, some lost, others stalled, and possibly a few "turned off." Is it as simple as that? Well, not really. The degree of correspondence between the actual performance and the expectation indicates only *what* is to be evaluated. You'll still need to consider *how* to evaluate achievement. Numerous techniques are available. Most types of evaluation, however, can be categorized into one of two modes: informal or formal.

Informal Evaluation

One kind of informal evaluation is the interview between teacher and student. This is a brief oral exam or "show me" request. The teacher can gain a great deal of insight about the student's understanding and capabilities with this method. The major problem is that it is time-consuming and favors the more verbal, unabashed type of student.

Informal *observation* can also be used for evaluative purposes. The purpose of observation should be directly related to the instructional goal. Results of such observations are usually organized in the form of anecdotal records. This method too can reveal pertinent information, but it requires a great deal of time for recordkeeping. Also, the evaluator must guard against the tendency to disperse a few inferences among the observations.

One of the more objective informal techniques is a *checklist* of expected behaviors. A rather thorough example of this method is presented in *Teacher Strategy Module: Self-Paced Investigations for Elementary Science* (Katagiri, Trojcak, and Brown, 1976, p. 25), a program for fifth- and sixth-graders as well as junior high school students (see Table 10-1). This format can be used as an ongoing means of assessment and modified to meet particular needs. It is comparable to the observation checklist and coding techniques presented in Chapter 8. This technique requires less time than interviews or observation and is more flexible and objective.

The final informal technique to be considered is *self-evaluation by the*

Table 10-1
CHECKLIST FOR EXPECTED BEHAVIORS*

STUDENT _____ MODULE _____										
PROCESS SKILLS	Identify									
	Describe									
	Demonstrate									
	Use rational numbers									
	Predict									
	Classify (order, distinguish)									
	Measure									
	Observe									
	Form hypotheses									
	Control variables									
	Analyze data									
	Define operationally									
	Design an experiment									
	Apply a technique									
	Develop a technique									
CONTENT	Grasps concepts and principles									
	Demonstrates learning									
RESPONSIBILITY	Completes investigations									
	Cares for and cleans up equipment									
	Completes record book									
	Exhibits humaneness									

*The vertical column heads at the top, which have been left blank here, would contain specific objectives or lessons, or even dates or grades. Blank lines within the table have been left so that the teacher can add other items. *Source:* George Katagiri, Doris Trojcak, and Douglas Brown: *Teacher Strategy Module: Self-Paced Investigations for Elementary Science,* Silver Burdett, Morristown, N.J., 1976. Used by permission of Silver Burdett Company.

Continued on page 300.

Table 10-1 (Continued)

STUDENT _____ MODULE _____												
SOCIAL SKILLS	Follows directions											
	Contributes to discussions											
	Works well with others											
ENTHUSIASM	Exhibits self-direction											
	Goes beyond requirements											

learner. This might be a checklist of behaviors, evaluated by the learner rather than the teacher, or an oral or written statement of how the learner feels about the progress or lack of progress being made. Again, time and lack of objectivity are possible problems with this technique. These problems are not as pronounced in the formal types of evaluation.

Formal Evaluation

Teachers tend to use formal types of evaluation more frequently than informal types. Probably no one goes through the educational system without having to demonstrate skills or understanding in some sort of pencil-and-paper tests. Numerous types of tests are possible, and many are available commercially—in fact, testing is becoming one of education's biggest enterprises. We'll consider several variations; again, you'll eventually have to decide if any of them fit your instructional objectives.

Situational and Performance Tests   In a *situational test* the students are asked to respond either orally or in writing to questions, or perform tasks, related to a teacher's demonstration or an explanation of a situation. In a *performance test*, the students (individually or as a class) are asked to demonstrate specified skills—for instance, the processes discussed in Part Two of this book. Students might also be asked to demonstrate their understanding by making sketches, constructing models, or giving examples of what was learned.

These evaluative techniques are excellent at providing direct evidence of skills and graphic or manipulative abilities. They give minimal emphasis to reading ability, which is a serious problem for practically all formal tests; and they reduce students' fear of not giving "the right answer." However, they are time-consuming; they require an abundance of materials if all the children are to be tested simultaneously; and they present real logistical problems. Consequently, many teachers usually adopt a more conventional version of formal tests.

Short-Answer Tests   This is one conventional form, and there's more variety in this type of test than in any other.

The *multiple-choice test* is probably one of its most common variations. A multiple-choice test is easy to score but rather difficult to construct and analyze. It is an efficient way to incorporate a vast amount of content, and it can be used over and over by simply varying the order of the items and changing some of the options. As in all short-answer tests, heavy emphasis is placed on reading ability; this can be unfair for evaluating learning in science. Multiple-choice tests also encourage guessing, and of course they give no opportunity for creativity. They measure the students' ability to recognize answers but provide no evidence that the students understand their choices.

The *completion test* has similar problems; and it perhaps goes even further than the multiple-choice test in encouraging memorization. This is also true of the *identification test* ("Select the correct answer from the words below . . .") and the *matching test* ("Match the terms in the right column with the explanations in the left column . . ."). But completion tests are quite effective if you're concerned about evaluating the students' retention of much material as quickly as possible.

Puzzles (especially crossword puzzles), *scrambled words,* and *picture tests* can lessen the fears and tensions usually associated with being tested. Some clever examples of these techniques in the form of thinking games can be found in De Vito and Krockover's *Creative Sciencing: Ideas and Activities for Teaching and Children (1976)*, and Abruscato and Hassard's *Whole Cosmos Catalog of Science Activities* (1977). Picture tests are most appropriate for very young children, since generally little or no reading is required. But these tests also encourage guessing, and they seldom measure higher levels of thinking. Also, their similarity to games can sometimes discourage children from putting forth their best efforts or encourage too much competition.

The *true-false test* seems never to die, even though it has been considered sick by both teachers and students. Constructing true-false items which are neither too simple nor too obscure is very difficult. This is certainly the easiest of all formal tests to score, but it is also the most conducive to guessing. Perhaps its greatest merit is that it usually provokes lively arguments after the "facts" are revealed, in which teacher and students have ample opportunities to improve their skills of persuasion (though generally no one really wins).

Long-Answer Tests   The two forms of long-answer tests most commonly used are *essay tests* and *oral reporting* or simply *orals*. (Because of the amount of writing required, essay tests are obviously inappropriate for young children who are still learning to write.) Both forms are relatively easy to administer but rather difficult and time-consuming to score. Even when a teacher has outlined the characteristics of acceptable answers in advance, it is difficult to avoid subjective assessment. But of all the tests, these offer the student the greatest opportunity for originality, synthesis, and the expression of personal understanding.

Designing and Using Evaluative Techniques

By now it should be clear that evaluation is not easy. But it is necessary and important to the entire teaching-learning process. Most students want to know, and have every right to know, what they've accomplished or failed to accomplish. The exception is probably the child who is still preoperational: if you are perception-prone and extremely egocentric, your way of viewing the world must be right; and if you are incapable of using logic and can think in only one direction, no explanations or check marks on tests will be convincing. But after the preoperational stage, that is, after self-conviction occurs, evaluation can sometimes provide a means of motivation, and not just for the successful student. It can contribute to that necessary aspect of learning, disequilibrium. This is true for you as a teacher, too—when your students "fail the test," you should look first at yourself.

The original answer to the question "How will we know we've arrived?" seemed relatively simple: refer back to the behavior specified in the instructional objectives. That's still the key. If your evaluation instrument identifies or measures the students' achievement of the instructional objective, your evaluation is fair and should provide valuable information. If you teach one thing and test for something else, your instrument is unfair. A *valid* test is an accurate evaluation instrument; it determines whether or not what was actually taught was actually learned. An invalid test deals with material which is unrelated to the instruction that took place. You should be concerned about the validity of any test: Does it measure what you want it to measure? (It's unreasonable to be evaluated on material one had no opportunity to learn.)

You should also be concerned about the ability of a test to withstand the effects of time—that is, a test should give consistent results when used again. The degree to which a test repeatedly accurately measures whatever is supposed to be measured is its *reliability*. If you are inexperienced in constructing evaluation instruments, you should be concerned first with the validity; after validity is established, you should look for evidence of reliability.

When you are considering the validity of test items (how well they match the requirements of an activity), you should also try to select the most *appropriate* type of evaluation. Even though situational and performance tests have their drawbacks, they're still the best means for determining the acquisition of skills. If what you wish to evaluate is the analysis and synthesis of information, a fill-in format is inappropriate. If your objectives included a variety of levels of understandings, guard against emphasizing just one level—especially the recall level. You might also weigh items according to their difficulty or importance. By doing this in advance, you will be better able to judge results impartially.

Design your directions and the actual test items as clearly as possible. Use the simplest words in clear sentences. In order to lessen the children's anxieties, begin with the easier items. If the children are to write answers, make drawings, or even write single letters or numerals directly on the test paper, allow ample space so that the children are not frustrated or tempted to bluff by writing answers on top of other words or illegibly small.

Avoid using direct quotes from curricular materials the students might be using. In general, avoid any items which require only memorization unless that was the extent of your objective. If at all possible, try to use a variety of evaluative items within one test, and occasionally vary the entire test format. If you have the time, energy, skill, and courage, try a more radical variation—let the children choose the mode of expression that best suits their interests and capabilities: written answers for those who prefer to write; drawings, sketches, or paintings for those graphically inclined; model construction for the mechanically skilled; body movement or dance; poetry; or simply an oral explanation.

If directions are necessary, be certain that they are clear to all the children. If the test format is unfamiliar to the group, provide one or two similar examples at the chalkboard or with an overhead projector. Allow ample time for the children to ask for clarification about the mechanics of the test.

Consider the best time of day for administering the test. Tests which require more time than a regular class period are probably too long.

Though this next suggestion is unpleasant, it must be considered. Try to reduce opportunities for cheating by separating the children as much as possible. Cheating is usually symptomatic of a much more serious problem or even a cluster of problems. It's doubtful that children are inherently dishonest. They seem to develop this trait from modeling the behaviors of those around them or to cope with pressure. Remember that you may be adding to this pressure by the way you emphasize and use (or abuse) evaluative procedures. It seems as though the majority of students from kindergarten through graduate school are obsessed with grades. Parents and teachers are often guilty of frightening children with their own obsessions. A constant diet of teaching and testing for the "one right answer" (especially if many of these answers deal with useless trivia) can drive just about anyone to cheating occasionally.

Try not to "teach for the test." If you are constantly asked, "Is this what we have to know?" or "What's on the test?" you should realize that you're on the wrong track. One way to overcome this tendency is to try to maintain a better balance between formal testing and informal evaluation. Let the students know that you also value their active participation in the learning encounters, their original questions, and any signs of original thoughts and actions.

Try to use any form of evaluation primarily as a learning experience. Discuss the tasks or items with the children after you have evaluated their performances. Help them to feel good about their successes and understand the reasons for their failures. Realize, and help your students realize, that evaluative techniques are meant to be *means*, not ends. They indicate achievement of goals, or difficulties in achieving goals. They help you determine your own skills in observing, predicting, distinguishing spatial and temporal relationships, measuring, organizing and interpreting data, and operationally defining your expectations and procedures. They help you communicate with the children's parents. And they will help you publicly defend what you value in terms of your teaching and the children's learning. Your values should mirror your actions; therefore, it's important that they be considered as part of the planning processes.

Are You Ready? Pause before Plunging

Most of the planning considerations discussed so far have dealt basically with factors within your control. The decision-making process also includes anticipating, being able to adjust plans when necessary, and coping with unexpected situations. Such skills are commonly referred to as the ability to "think on your feet." It's impossible to be totally prepared for the unexpected. But perhaps by considering some possible examples, you'll be better able to respond appropriately to unexpected events or surprising conditions.

Anticipating Possible Problems

What are some probable sources of problems which could hinder instructional planning? In your classroom observing experiences, what problems occurred? Which of these were seemingly unanticipated? Was there any evidence that anticipated problems occurred? Listing specific sources of possible problems could become an endless task; instead, let's consider four major categories.

Students One child or several children may come to class with emotional problems. Such problems might vary from fleeting disappointments (like having been beaten in a game of checkers the night before) to very frightening and disturbing experiences (like having seen a drunken father beat up a screaming mother). Children's sudden illnesses can disrupt lessons, as can serious accidents in the classroom or on the playground. Children's going to the bathroom should be an insignificant matter, but it can be disruptive if handled poorly—remember that a quick exit causes little distraction, but an unsuccessful attempt to leave can cause a great deal. A lesson can also be hindered if the children are fatigued or hyperactive. It's a mistake to expect children to "settle down" immediately after recess, for example. Fights, either verbal or physical, eruptions by individual or small groups, and even classroom pranks can hinder your lesson. It is also difficult to carry out a plan when the students are anticipating something more enjoyable—like a school party or an approaching vacation—or something out of the ordinary. Unexpected changes in the weather can also distract the students and hinder your plans. Sometimes it is very difficult to detect until after the fact, but even students' previous failures or frustrations with the topic you are presenting can elicit unanticipated problems.

Disruptions Announced or unannounced, disruptions can foil your plans. There's nothing quite like a loud voice over the intercom to shatter a lesson on listening to natural sounds outside the classroom windows. Even anticipated disruptions such as fire drills, storm drills, picture taking, testing programs, and assemblies have effects which last longer than the events themselves. The arrival of unexpected visitors—or the unexpected absence of expected visitors—can cause problems and create a need for quick adjustments. The presence of certain objects—such as the arrival of a fire truck for the kindergarten, or a child's pet python for "show and tell"—can send disruptive vibrations throughout the entire school. If a nearby classroom is especially noisy, that too can be disruptive. Breakdowns of the physical plant or equipment also create disruptions—falling plaster, flickering of fluorescent lights that need replacing, faulty plumbing, frozen pipes, an ill-functioning furnace—as does the coming and going of maintenance workers.

Instructional Materials Instructional materials can be damaged or ruined in shipment, and if they are not unpacked until the last minute, this can create a real problem. Live specimens can be hale and hearty when left at the end of a school day and sick, dead, or mysteriously gone by the next morning's lesson. (Many unmarked brine shrimp, tadpoles, and other aquatic cultures have met ignominious fates at the hands of overtidy janitors.) Because of poor communication or lack of cooperation, materials that were supposed to be shared may not be available. Materials that were to be collected by the children can be forgotten or lost. Unexpected changes in the weather might make it impossible to use certain materials. Sometimes a teacher may discover that seemingly attractive materials are considered unappealing by the children; or the opposite may occur—apparently bland and mundane items might, unexpectedly, over-stimulate the children. One of the most difficult problems with the use of materials can simply be categorized as "unknown": "It worked at home when I tried it; I don't know why it fizzled at school."

The Teacher Teachers can unconsciously be the source of many planning problems. As with the students, emotional and physical difficulties can arise. Teachers, too, become impatient, irritable, moody, physically ill, and tired. Uncertainty or lack of confidence can also be a problem, since children can often sense it. The more unclear the goal or the directions for the lesson, the greater the potential for problems and even for panic. But overplanning—or at least overcommitment to plans—can also cause troubles. The inexperienced teacher sometimes has the unconscious attitude: "I've put so much time and effort into this plan that we're going through with it—come what may." When this happens, the plan becomes an end in itself rather than a means. It is true, though, that most problems seem to stem from being underprepared rather than overprepared. Insufficient planning often results in an inability to anticipate problems, which is itself a serious problem—exceeded, perhaps, only by unawareness that any difficulties exist.

Personalizing the Basic Questions

Meaningful and vital learning is interiorized learning. That which is learned becomes a part of you after you have given some part of yourself to the learning process. Certainly, the more you give of yourself—that is, the more fully you involve your total being in the learning process—the more your perception becomes your understanding. What makes it possible for you to give of yourself in meaningful ways? I believe that you must first *know yourself* better.

Understanding, acceptance, and love of children and the natural world must begin with understanding, acceptance, and love of yourself. In other words, "You can't give what you haven't got." It's extraordinarily difficult, if not impossible, to handle the many variables involved in planning and presenting science with children when you can't accept or—worse yet—even recognize your true self. This chapter cannot *cause* you to plan and eventually implement successful science encounters for children; nor can this final section *cause* you to know yourself better. Both these things can happen only within you, as results of your own actions. "Personalizing" the four basic questions of this chapter may be a good way to begin.

Instead of asking:	*Try considering:*
Where are we?	Where am I? Who am I?
Where are we headed?	Where am I headed as a science educator?
How do we get there?	How do I get there?
How will we know we've arrived?	How will I know I've arrived?

These are tough questions. Definitive answers are probably impossible, since change is such a constant reality in all our lives. But they're certainly questions worth pursuing if you hope to gain more insight into yourself. The search for answers will also help you develop a more personal understanding of science in the real world of the classroom.

Where Am I? Who Am I?   It is difficult to identify and define the *real* person in each of us. It is far more important and realistic to discuss *becoming* a person or simply the process of becoming. The search for selfhood has probably never been more popularized than in recent times. There is an ever-increasing number of methods, techniques, books, and programs on self-awareness, self-development, and sensitivity. One can choose from encounter sessions, transactional analysis, mind control, yoga, synectics, values clarification, gestalt therapy, biofeedback, body movement, confluent education, transactional meditation, psychosynthesis, and prayer groups. But it's doubtful that there is any foolproof program, or any shortcut, for finding, accepting, yet always inproving your real self.

In the broadest possible terms, you are the total collection of your thoughts, feelings, and actions. What kind of balance do you generally maintain among these three factors? What evidence do you have that one might outweigh the others? If you like mental imagery, envision a teacher who is "all heart" but never thinks or acts; or one who is "all brain," with forty-eight graduate credits in science, but is inert and apathetic; or a perpetual-motion machine who is too busy to think or feel. Your head, heart, and hands must work in coordination as you think, feel, and act. In order to do science with children well, all three factors must be present and united. When you examine your instructional plans, be sure that you have made provisions for the children to demonstrate these three components also.

What words would you use to describe where you are in your own development, and who you are? Do you think these would be basically the same words others would use to describe you? You are the only you in all the world—does this please you or disturb you? If you are not accepting of yourself, it is unlikely that you can be receptive to or attentive of others, or even of objects and events around you. Self-acceptance and self-esteem affect your learning just as they affect children's learning—perhaps even more. If you feel inadequate or "boxed in," it's very difficult to give of yourself freely, to risk, or to make new discoveries. Children are often better able to take risks and make discoveries than adults are, having not yet lost their sense of wonder or their enthusiasm for using their five senses. For children, the avoidance mechanism "What will people think?" is virtually nonexistent. Their openness to all life and living is what you need to rediscover if you are ever to be ready and able to do science with them.

Where Am I Headed as a Science Educator?   We can never just "be": it's impossible to stand still or achieve a permanent state of equilibrium. We must gradually and continuously become the unique persons we can become. This becoming process will depend primarily on what goals we select for ourselves. No one can (or should) mandate goals for anyone else. The goals must be free choices of free persons.

What does it really mean to be a free person? Basically, a free person has the inward strength and convictions to make decisions based on *self*-determination. You will not hear a free person making rationalizations like, "The system's against me," or "I've got to conform or I'll be out," or "Fight fire with fire" or "I'm a Capricorn and my horoscope for today said that I should . . ." A free person makes free, deliberate choices and is not dominated by outside forces. A free person is usually restless without being nervous about such restlessness. He or she is always searching, is never complacent, and resists being bound to the status quo. A free person will never say, "That's the way it's always been, so that's the way it has to be." A person who is free has a propensity for discoveries and is able to find joy in even the smallest discoveries. This person can laugh at honest mistakes, can bounce back, can play both physically and mentally, can take risks, and can even fail without collapsing. As a full participant in the process of becoming, the free person can *realistically* say, "I can become whatever or whoever I want to become." Most important, *the free person is willing and able to allow others to be free.* This last characteristic is by far the most difficult to acquire; but it is the real hallmark of a truly free person, the sign of trust in others and confidence in oneself—the ability to let go without the fear of loss, to give without expecting a return.

To what extent are you free to become the kind of science educator you want to be? To what extent are you not free? How might you become freer or more consciously self-directed toward where you want to go as a science educator? As you ponder these questions, be honest with yourself—but be gentle, too. Center on your positive qualities rather than your weaknesses. Realize that it's basically your attitude that enables you to say, "I can" instead of "I can't," or "I will" instead of "I won't." Know, too, that disequilibrium is a vital part of becoming.

How Do I Get There?   It's important to speculate about the future and to learn from past experiences as you continue becoming the kind of science educator you want to be. But you also need to capitalize on the reality of the present, that is, the *here and now*. It's the present that is the basic substance of both your being and your becoming. How often do you get caught up in fretting about the past and worrying unnecessarily about the future, ignoring the fullness of the present? Recognizing the here and now and responding to it are the most basic steps in your "task analysis" of your personal goals. They are also important prerequisites for developing awareness in general.

If you lack awareness, it's also likely that you'll lack responsiveness—or what Fritz Perls calls "response-ability" (1969, p. 65). Throughout this entire book you've been asked to be responsive to children, to the thinking operations, and to some of the important instructional skills. Your responsiveness will determine many of your decisions—for example, self-directed learning as opposed to "This is what should be memorized"; active involvement with materials as opposed to "I can't stand a messy science lesson"; personally planning science materials for your children as opposed to "Just follow what's in the teacher's guide."

You should also check your responsiveness to one of the most basic questions that can be asked: "Why do you want to be an educator (or teacher)? What is your most honest answer? Is it a matter of building your ego by being in command, having power, knowing the answers, and directing a captive audience? Is it because you truly value learning, want to help children learn, and want to be a continuous learner yourself? Or does the reason lie elsewhere? For instance, you may never have really considered other alternatives; or you may have considered education the easiest choice; or you may have drifted into education and then found that changing was too much trouble. However you came to make your choice, are you pleased with it? Even if you can't answer that, your students probably can—they have an uncanny ability to sense whether or not you value what you're doing.

Your responsiveness is demonstrated by your sensitivity to the entire process of teaching and learning. Even after you've completed the most thorough planning, once you present the science encounter to your children countless new variables may appear. At that point, how responsive will you be? To what extent can you be flexible, patient, and able and willing to try alternative procedures? Will you begin, proceed with, and run through the science lesson like a bulldozer in high gear? Will you invite, or will you command? Will you lead, or will you force? *Will you care*—about the children, about what they are expected to do and experience, about whether you can share yourself, about how you value what is to be learned, and about the worth of the entire interactive learning process? In *What Do You Say to a Child When You Meet a Flower?* David O'Neill makes the following suggestion:

Now you may, if you wish, share your adult learning with the child. You could give your simple lesson in biology and botany and ecology and economics. But don't try to do this before you have met the child and met the flower. Who is interested in the price of roses in New York if he has never met a rose? So be sure to meet the rose first, and to meet the child. Then, if you do have something good to say about flowers, it will be really you talking and not someone just pretending to be you. If the child has enjoyed meeting you and meeting the flower, he may be interested in what more you have to say (1972, p. 23).

How Will I Know I've Arrived?   For once, a definitive answer is possible: As long as you keep trying to become a more effective science educator, you'll never arrive completely or finally. You are constantly changing, and so are children, the body of science that might be presented, the world we live in, and the needs of the times. The best you can do to answer this question is to approximate relative progress.

The general method for determining your personal progress is similar to evaluating children's learning. You must establish your personal objectives for self-improvement. As with instructional objectives, the more specific, observable, and measurable your personal goals are, the more easily you can tell whether you've reached them. If you wake up each morning and say, "Today I'm going to do better at science with children," you might as well roll over and continue sleeping and dreaming. You'll need to be far more specific if you sincerely want to plan for and pursue self-improvement.

It is helpful to run a mental playback at the end of each science encounter with children. You might ask yourself the following questions:

What went well today? How and why did I succeed?
What went poorly today? How and why did I not succeed?
What changes should be considered for tomorrow?
Were I to present this science lesson again, what would I do differently?

By continuing to ask yourself these questions over a period of time, you will eventually discover some revealing patterns, probably both positive and negative. These patterns could become the bases for the specific personal goals needed for honest self-confrontation and accurate self-evaluation.

Some Final Considerations

As you acquire and eventually test your decision-making skills, you must realize that you alone are responsible for making yourself the science educator you want to become. You may take suggestions, advice, and ideas for techniques from others; but you alone must make the ultimate decisions. The person you are, the person you are becoming, and your classroom decisions will, of course, greatly affect your science with children. However, you will not bring about their learning: you cannot force learning upon them, any more than improvement can be forced upon you.

Whether or not you are consciously willing to respond to the final four questions—Where am I and who am I? Where am I headed as a science educator? How do I get there? How will I know I've arrived?—the questions are being answered. Noel McInnis has defined an *environment* as anything or anyone which is affecting or is being affected by something or someone (1972, p. 48). If you accept this definition, you must also accept the fact that each of us is an environment, since each of us is always affecting or being affected by something or someone. With this realization comes responsibility: our influences will continue—but how? The answers reside in additional questions:

As an environment, what do you influence? What influences you?

As an environment, what do you do for and with others? What do others do for and with you?

As an environment, what do you contribute to others' becoming? What do others contribute to your becoming?

Biologists have identified specific signs as characteristic of certain environments. If you were to wear a sign that characterized *you* as an environment, what would it say?

I began this book with some personal comments. I'd like to conclude by making a few more. I've often asked myself the same questions that I have been posing for you. I know how difficult it is to answer them and to make the answers part of my daily life. I strongly believe that the struggle to become the best person possible is worth whatever it costs—that the people and situations I encounter deserve my responsiveness. In the Introduction, I expressed the hope that your interactions with this book would result in both personal and professional growth that would help you improve your ability to do science with children. I also hope that you have the courage to make thoughtful decisions about science with children, about being and becoming, and about recognizing human potential as fully as possible.

Encounters

1. Practice constructing some instructional objectives related to any of the activities you completed in Part Two. Identify the four major components: learning condition, performance terms, standards for acceptable performance, and rationale. Compare your products with those of others. Which components were the most similar? Which varied the most?
2. Analyze and evaluate the objectives present in some samples of elementary school science curricula. Which component was most obvious? Which was least obvious? How might these objectives be improved?
3. A strong case was made for including a rationale in instructional objectives. On the basis of your personal experiences either as a student in science classes at any level or as an observer of science being taught to others, explain (a) the use, absence, or abuse of the rationale and (b) the implications for the learner, the teacher, and the science curriculum.

4. What do you consider to be the most significant advantages and disadvantages of instructional objectives? Interview some elementary school teachers and analyze their opinions about instructional objectives. Compare the various viewpoints.

5. What factors would you want to consider when deciding whether to present a science lesson to the entire class or to small groups of children?

6. If possible, spend some time in a classroom observing and analyzing how elementary teachers go about selecting materials for science lessons, organizing and distributing these materials, and storing them. In what ways did these decisions help or hinder the children's learning or attainment of the instructional objectives? Had you been presenting the lessons, what decisions might you have made?

7. Numerous examples were given of problems which could disrupt your instructional plans. Which do you think would upset your plans the most? Why? How might you avoid or lessen the effects of such problems?

8. Design a plan for presenting a specific science encounter for children. Which components of your lesson plan are most important? Then implement your plan with children in an actual classroom. Evaluate the results: What went well? What went poorly? If you were to present it again, what changes would you make?

9. Plan a sequence for a minimum of five related science encounters for children. Select an appropriate terminal goal and develop a motivational technique which makes the goal worth achieving. Diagram the sequence of related objectives by means of a task analysis. If possible, implement your mini-unit and evaluate the results. Also, analyze your decision-making skills; which were the most important?

10. Review the four major questions in the section "Pause before Plunging." What is your idea of an effective science educator? Which characteristics of an effective science educator do you have? Which do you lack? Decide on some specific goals worth striving toward which will help you become more like the person you described.

11. Your choice.

References

Abruscato, Joe, and Jack Hassard: *The Whole Cosmos Catalog of Science Activities*, Goodyear, Santa Monica, Calif., 1977.

De Vito, Alfred, and Gerald H. Krockover: *Creative Sciencing: Ideas and Activities for Teachers and Children,* Little, Brown, Boston, Mass., 1976.

Katagiri, George, Doris Trojcak, and Douglas Brown: *Teacher Strategy Module: Self-Paced Investigations for Elementary Science,* Silver Burdett, Morristown, N.J., 1976.

Mager, Robert F.: *Preparing Instructional Objectives*, Fearon, Belmont, Calif., 1962.

———: *Goal Analysis*, Fearon, Belmont, Calif., 1972.

McInnis, Noel: *You Are an Environment*, Center for Curriculum Design, Evanston, Ill., 1972.

O'Neill, David P.: *What Do You Say to a Child When You Meet a Flower?* Abbey Press, St. Meinrad, Ind., 1972.

Perls, Frederick: *Gestalt Therapy Verbatim*, Real People Press, Lafayette, Calif., 1969.

Suggested Readings

Bybee, Rodger W.: *Personalizing Science Teaching*, National Science Teachers Association, Washington, D.C., 1974. (A brief monograph on the important personal roles a teacher has in facilitating children's learning science.)

Castillo, Gloria A.: *Left-Handed Teaching: Lessons in Affective Education,* 2d ed., Praeger, New York, 1978. (A model for confluent education is presented, consisting of four factors: cognitive domain, affective domain, readiness-awareness, and responsibility. The bulk of the book consists of 211 affective lessons.)

Curwin, Richard L., and Barbara Schneider Fuhrmann: *Discovering Your Teaching Self: Humanistic Approaches to Effective Teaching,* Prentice-Hall, Englewood Cliffs, N.J., 1975. (This poses some searching questions related to teaching and learning. Numerous examples and suggestions are presented, but you must ultimately develop your own answers.)

Holt, John: *What Do I Do Monday?* Dell, New York, 1970. (Holt gives a very personal account of his strong belief that children learn by doing and that self-esteem is the best catalyst for learning in general. Approximately half of the book concerns various styles of teaching.)

Mager, Robert F.: *Developing Attitude toward Learning*, Pearon, Palo Alto, Calif., 1968. (An extremely practical paperback on how to plan instruction as well as how to recognize students' attitudes toward learning, either pro or con.)

Maslow, Abraham H.: *The Farther Reaches of Human Nature*, Viking, New York, 1971. (This is the apex of Maslow's works and his rich contributions to human psychology and the process of becoming a self-actualized, fully authentic human being.)

Miller, John P.: *Humanizing the Classroom*, Praeger, New York, 1976. (A firm foundation is presented from models dealing with personal development, identification of self-concepts, sensitivity training, and expansion of consciousness. Numerous classroom applications are also suggested.)

Raths, Louis E., Merrill Harmin, and Sidney B. Simon: *Values and Teaching*, Charles E. Merrill, Columbus, Ohio, 1966. (The theoretical nature of values and valuing is clearly presented, along with the value-clarifying method. The techniques can benefit both you and the students as regards determining what you value and even who you are and where you are.)

Rogers, Carl R.: *Freedom to Learn*, Charles E. Merrill, Columbus, Ohio, 1969. (Rogers deals with creating a climate of freedom—both in oneself and in the classroom—in which becoming, learning, and valuing can occur more fully.)

Romey, Williams D.: *Risk-Trust-Love: Learning in a Human Environment*, Charles E. Merrill, Columbus, Ohio, 1972. (Suggestions are offered for developing a more trusting learning environment in both elementary and secondary schools and for reexamining one's role as a teacher.)

Simon, Sidney B., and James A. Bellanca (eds.): *Degrading the Grading Myths: A Primer of Alternatives to Grades and Marks*, Association for Supervision and Curriculum Development, Washington, D.C., 1976. (This is a valuable collection of nineteen articles dealing with the problems of grading, some research findings, five alternative procedures, and suggestions for changing the grading system.)

Sund, Robert B., and Rodger W. Bybee (eds.): *Becoming a Better Elementary Science Teacher*, Charles E. Merrill, Columbus, Ohio, 1973. (Although all thirty-six articles included in this book of readings are good, those dealing with becoming a better science teacher through understanding self—Section I—and through understanding the facilitation of learning—Section IV—best supplement this chapter.)

Thatcher, David A.: *Teaching, Loving, and Self-Directed Learning*, Goodyear, Pacific Palisades, Calif., 1974. (Six modes of teaching are considered which increase the child's self-esteem, the desire to learn, and opportunities for independent learning.)

Trojcak, Doris A: "Implementing the Competency of Sequencing Instruction," in James E. Weigand (ed.), *Implementing Teacher Competencies*, Prentice-Hall, Englewood Cliffs, N.J., 1977. (This chapter will provide additional information and examples for applying task analysis. The checklist for evaluating your sequenced instruction on pp. 215–216 is quite useful for any type of instructional planning.)

Zaharik, John A., and Dale L. Brubaker: *Toward More Humanistic Instruction*, Brown, Dubuque, Iowa, 1972. (The authors establish a firm philosophical foundation for humanistic education from which teachers can build the technical dimensions of decision making. Both descriptive and prescriptive content is offered—that is, what is and what should be.)

INDEX

315